THE ESSENCE OF
ELECTRIC POWER SYSTEMS

THE ESSENCE OF ENGINEERING SERIES

Published titles

Forthcoming titles

THE ESSENCE OF

ELECTRIC POWER SYSTEMS

J. A. Harrison B.Sc., Ph.D.
Lecturer in Electrical Engineering and Electronics
University of Liverpool

Prentice Hall

LONDON   NEW YORK   TORONTO   SYDNEY   TOKYO
SINGAPORE   MADRID   MEXICO CITY   MUNICH

First published 1996 by
Prentice Hall Europe
Campus 400, Maylands Avenue
Hemel Hempstead
Hertfordshire HP2 7EZ
A division of
Simon & Schuster International Group

Typeset in 11/13pt Times
by MHL Typesetting Ltd, Coventry

Printed and bound in Great Britain by
T. J. Press (Padstow) Ltd

Library of Congress Cataloging-in-Publication Data

Harrison, J. A., Ph.D.
    The essence of electric power systems / J.A. Harrison.
       p. cm. — (Essence of engineering)
    Includes index.
    ISBN 0-13-397514-2
    1. Electric power systems. I. Title. II. Series.
  TK1001.H382 1996
  621.31–dc20   95-33548
                                      CIP

British Library Cataloguing in Publication Data

A catalogue record for this book is available from the British
Library

ISBN 0-13-397514-2

---

1 2 3 4 5   00 99 98 97 96

To Sonja, Deborah, Julian and Oliver

Contents

Preface

This book is based on a short introductory course given to students in the Department of Electrical Engineering and Electronics at the University of Liverpool. Some of these students later specialized in power systems, and for them the course was intended to lay a solid foundation on which their further studies could be built. For other students the course was the sum total of their formal teaching in power systems. The book does, I hope, provide a balanced view of the main features of electric power systems, together with some of the problems facing the power engineer. I have tried also to answer in the text a number of questions about power systems that puzzle many people who are not electrical engineers. For example, why do we have unsightly pylons in the countryside when electricity can be carried underground in cables? Why do the pylons have to be so large? What is the purpose of a pumped storage scheme? I have assumed that the reader will have a knowledge of single-phase electrical circuits; three-phase circuits are introduced in the text. The mathematics required is not very demanding: school sixth-form standard would suffice.

Why another book on power systems? There are many very good, comprehensive and detailed textbooks covering the subject, but I have found, from talking to students, that many of them are unwilling to buy such books. This is because they go into the subject in far too much detail which can be bewildering to a student meeting the subject for the first time, and also because these books are relatively too expensive for a student who may only be taking the subject at an introductory level.

I hope you will find this book readable, easy to understand and perhaps stimulating so that you will go on to study power systems at a more advanced level.

Tony Harrison

Acknowledgements

The material in this book has been gathered together over a long period from many different sources, but I am specially grateful to the following: the Institution of Electrical Engineers and the authors R.J.S. Ward and M.N. Eggleton for extracts from 'Electrical parameters of 400 kV and 275 kV overhead lines in England and Wales' in IEE Conference Paper No. 44, September 1968, to compile Table 4.2; Guinness Publishing Ltd for data on the power station using direct sunlight, taken from the *Guinness Book of Records*, 1995; the National Grid Company for Table 7.3 and other data; BICC Cables Ltd for information on high-voltage cables and particularly N.R. Garnet and J. Vail for help in interpreting this data.

Symbols

SI units (Le Système International d'Unités)

A	ampere, the unit of current
F	farad, the unit of capacitance
H	henry, the unit of inductance
Hz	hertz, the unit of frequency, i.e. the number of cycles per second
J	joule, the unit of energy
K	kelvin, the unit of absolute temperature
m	metre, the unit of length
rad	radian, the unit of plane angle
s	second, the unit of time
T	tesla, the unit of magnetic flux density
V	volt, the unit of potential difference and electromotive force
W	watt, the unit of power
Wb	weber, the unit of magnetic flux
Ω	ohm, the unit of resistance, reactance and impedance

SI prefixes

p	pico, 10^{-12}
n	nano, 10^{-9}
μ	micro, 10^{-6}
m	milli, 10^{-3}
c	centi, 10^{-2}
k	kilo, 10^{3}
M	mega, 10^{6}
G	giga, 10^{9}
T	tera, 10^{12}

Quantity symbols

A	area
B	magnetic flux density
C	capacitance
d	distance
d_{eq}	equivalent distance
E	root mean square value of the electromotive force
e	instantaneous value of the electromotive force
e_0	peak value of the electromotive force
f	frequency in Hz
h	height
I	current
I_B	base value of the current
I_{pu}	per-unit value of the current
I_{sc}	short-circuit current
J	polar moment of inertia
j	square root of minus one
L	inductance
N	number of turns, or transformation ratio
P	real power
Q	reactive power
R	resistance
r	radius
S	apparent power
S_B	base value of the apparent power
S_{sc}	short-circuit apparent power
t	time
V	root mean square value of the potential difference (voltage)
V_B	base value of the voltage
V_L	line voltage
V_p	phase voltage
V_{pu}	per-unit value of the voltage
V_T	Thévenin equivalent phase voltage
v	instantaneous value of the voltage
X	reactance
X'	transient reactance
X_{pu}	per-unit value of the reactance
X_s	synchronous reactance
Z	impedance

Z_B	base value of the impedance
Z_{pu}	per-unit value of the impedance
ΔV	small voltage difference
δ	torque angle, load angle or transmission angle
μ	permeability
μ_0	permeability of free space
Φ	magnetic flux linkage
ϕ	phase angle
$\cos \phi$	power factor
ω	angular frequency

Abbreviations

a.c.	alternating current
d.c.	direct current
e.m.f.	electromotive force
ln	natural logarithm
pu	per unit
rev/min	revolutions per minute
r.m.s.	root mean square
V Ar	volt ampere reactive (var)
°C	degree Celsius

Bold italic type denotes the quantity is a phasor.

CHAPTER 1
Introduction

In this chapter we shall begin by understanding what is meant by an electric power system and briefly consider the advantages of electrical energy, particularly in alternating current form. Then we look at the different companies which are responsible for the generation, transmission and distribution of electricity in the United Kingdom. Next we look at the growth of the electricity supply industry and its impact on the environment. Finally we consider some recent trends in the use of different energy sources for electric power generation.

We may well begin by asking, 'what is an electric power system?' An electric power system, or as it is sometimes called today, an electric energy system, is the name given to a group of power stations, transformers, switchgear and other components which are interconnected by overhead lines and underground cables, to supply consumers with electricity. Power systems developed many years later than power generation. In the early years each power station supplied its own local load. Some generators produced alternating current (a.c.) and others direct current (d.c.). There was soon rivalry between the advocates of a.c. and the advocates of d.c., but a.c. quickly prevailed. There was then a steady development of generating stations until each large town or load centre had its own station.

The advantages of alternating current are easy to appreciate. The outstanding advantage is that the sinusoidal voltage can be stepped up or down by using a transformer. This means that each part of a power system can operate at its optimum voltage. Thus the generators (alternators) can produce power at around 11 000 V to 22 000 V (11 kV to 22 kV) to suit the designer. The loads can take power at 230 V, 400 V, etc., as required by the consumer, and the transmission of power can take place at 132 kV, 275 kV, 400 kV or even higher voltages. These very high voltages are essential for the efficient transmission of electric power. This is because, for a given power, as the voltage is transformed up, the current is transformed down by the same factor. Thus a 400 kV line carries only one-third as much current as a 132 kV line, if they are transmitting the same power. Power losses in the lines are mainly due to resistive heating of the conductors which is proportional to I^2R (current squared times resistance) so that, for the same power and the same size of conductor, the losses in the 400 kV line would be only about one-ninth of those in the 132 kV line. Another advantage of alternating current compared with direct current is that it is much easier to design and construct circuit breakers to interrupt alternating current. The reason for this is explained in Chapter 4. In the United Kingdom and the continent of Europe the system frequency is 50 Hz, but in the American continent it is 60 Hz.

Electricity is not itself a source of primary energy, but it is by far the most versatile and convenient form into which the primary energy of coal and oil, nuclear energy and the potential energy of stored water can be transformed. Electricity can be converted into heat with 100 per cent efficiency and into mechanical motion with very high efficiency. This is done without producing any pollution or waste product of any kind. Electricity is also the only possible source of power for a wide range of electronic goods.

For any equipment, electricity generated in power stations is a much cheaper source of energy than primary batteries (dry cells). Secondary storage batteries offer a convenient way of making this energy available to portable equipment. If even better secondary batteries could be developed they would provide a very attractive way of powering small motor vehicles, particularly private cars. What is needed is a battery which is lighter, cheaper and longer lasting than present commercial batteries. The cost of running a car on electricity as compared with the cost of running it on petrol (gasoline) depends on the various efficiencies involved as well as the current prices of the fuels, but the cost of electricity to run a car is likely to be much less than the cost of petrol. There would be an additional advantage, if cars were charged overnight, in that the charging load would help to improve the power system load factor. The meaning of the system load factor is explained in Chapter 7.

In England and Wales electrical energy is generated by independent companies, two of which are National Power and PowerGen. A distinction is made between transmission and distribution of electricity. Transmission is the bulk movement of energy from power stations to load centres, and from one load centre to another. It is the responsibility of the National Grid Company and takes place at 400 kV and 275 kV. Distribution is the movement of electrical energy from substations to consumers, and takes place at 132 kV, 66 kV, 33 kV, 11 kV, etc., down to 400 V three phase and 230 V single phase. At privatization of the electricity industry, 12 regional electricity companies (London Electricity, Manweb, Northern Electric, etc.) were set up with responsibility for distribution. Scotland has Scottish Power and Scottish Hydro-Electric which are integrated generation and supply companies. The situation with nuclear power stations in the United Kingdom is that the older ones will remain in the state sector while the more modern ones are privatized. The whole transmission system in England, Wales and Scotland is interconnected and synchronized. In Northern Ireland the generating stations are owned by several different companies but Northern Ireland Electricity is responsible for the transmission, distribution and supply of electricity in the province.

A rapid growth in the amount of electricity generated in industrial countries took place after the Second World War. In the USA, for example, during the period 1945 to 1970 generating capacity doubled approximately every 10 years. This represents an average increase each year of about 7 per cent on the previous year's total. However, during the 1970s the steady growth of the electricity industry faltered, mainly because of large increases in the price of oil especially, and coal to a lesser extent. This affected the demand for electricity in three ways. First, it caused a recession in industrial activity which was reflected in a reduced demand for electrical energy. Second, it made many organizations energy conscious so that they made efforts to save all forms of energy, electricity included. Third, it made electricity generated conventionally from coal and oil less competitive than rival energy sources, particularly natural gas, because the cost of electricity tends to reflect the cost of the primary fuels from which it is generated.

Because of their very large size, power stations have a considerable impact on the environment. This takes a number of forms. First, most modern coal, oil and nuclear power stations are very large and visible over a wide area, particularly if they have a chimney and cooling towers. A high chimney is necessary, when burning coal or oil, to dissipate the flue gases well away from the neighbourhood of the station and ensure that they are sufficiently diluted by atmospheric air before they reach ground level. Some critics claim that this merely transfers the pollution to another area or even to another country. There have been claims that acidic rain in Scandinavia, originating from the sulfur dioxide from tall chimneys in the United Kingdom, has inhibited the growth of trees in Norway and Sweden. The sulfur dioxide is produced when coal or oil containing sulfur is burnt in power stations. It is possible to remove the sulfur dioxide from the flue gases by adding flue gas desulfurization units, but these increase both the capital and the running costs of the power stations. Sometimes the large plumes which hang over power station cooling towers are mistaken for pollution. They are in fact nothing more than water vapour and can be looked upon as artificial clouds. Some environmentalists worry about the hot water discharged by thermal power stations. This can indeed upset the balance of life in a river, but in many cases it can provide conditions in which fish thrive. There could well be a case for using this warm water for fish farming on a large scale. Another possible use is in horticulture where the warm water can be used to heat greenhouses. In some countries the waste heat is used on a large scale for district heating of blocks of flats but this is associated with a reduction in the steam cycle efficiency of electricity generation.

Another aspect of modern power systems which has a considerable impact on the environment is the presence of the overhead power transmission system. With increases in power-line voltages there has of necessity been an increase in the size of the towers or pylons which support the lines. Increases have also been made in the size and number of the conductors. Both of these factors make modern overhead lines more obtrusive. It is technically possible to put all the transmission system below ground using high-voltage cables. The reasons why this is not done are explained in Chapter 4.

Recently, nuclear power for electricity generation has become unpopular. This is in part due to worries about safety following the Chernobyl accident and also concern about the very high estimates of decommissioning costs. In the very long term, when fossil fuels are seriously depleted, nuclear and solar energy are likely to provide the main sources of primary energy for electricity generation. Already solar energy in the form of wind-powered generators is making a small contribution.

New power station developments have moved away from the very large coal- and oil-burning plants, in favour of natural-gas-fired generating systems. High efficiencies are obtained by using combined-cycle gas turbine plants and also combined heat and power plants.

1.1 Summary

A power system consists of generators to produce electricity, and transformers, switchgear and other components which are interconnected by underground cables and overhead lines to supply it to consumers. The overhead power lines use high voltages for efficient transmission of electrical energy. There are many companies involved in electricity generation, transmission and distribution in the UK. The power systems they operate have a considerable impact on the environment, both visually and because some power stations emit noxious flue gases. Recent power station development has favoured natural-gas-fired plants rather than coal, oil or nuclear power stations.

CHAPTER 2

Three-phase systems

In this chapter we shall begin by explaining why power systems nearly always use a three-phase arrangement for electricity generation, transmission and distribution. We shall study the basic three-phase generator, star and delta connections, and the relations between phase and line quantities. Finally we shall see the advantages of interconnecting generators which lead to the development of a grid system.

2.1 Advantages of three phase

While domestic consumers nearly all take their electricity in the form of single-phase power, all large power systems use three phases. There are a number of advantages in having a three-phase system: (i) power and torque are constant in a three-phase motor or generator, which means that the machines will run more smoothly than single-phase machines in which the torque pulsates at twice the system frequency; (ii) for a given size, i.e. for a given amount of steel and copper, a three-phase machine has a greater output power; (iii) a three-phase transmission system will carry a greater power than a single-phase system, other things being equal. The reason for this will be explained later. The second and third points are really economic considerations, and because power systems are so large and costly, economic considerations influence this and many other aspects of power systems.

> **Tutorial question 2.1**
>
> Give the expression for the total power loss in a three-phase overhead line which has a resistance of R ohms per phase and carries a current of I amps.
>
> $(3I^2R \text{ watts})$

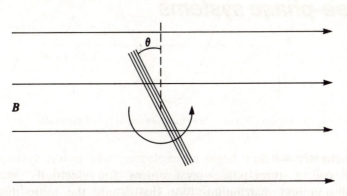

Figure 2.1 *Simple single-phase generator*

2.2 **Star and delta connections**

To help to understand a three-phase generator, consider first a simple single-phase generator consisting of a narrow coil of N turns and area A square metres, rotating in a uniform magnetic field of flux density B tesla, as shown in Figure 2.1. With the coil in the position shown in Figure 2.1, the magnetic flux, Φ, linked by the coil is $NAB \cos \theta$ weber. When the coil is rotating, $\theta = \omega t$, where ω is the angular frequency of rotation of the coil in radians per second, and t is the time in seconds. Therefore the magnetic flux linkage is given by:

$$\Phi = NAB \cos \omega t$$

By Faraday's law the induced electromotive force (e.m.f.), e, is equal to minus the rate of change of flux linkage, i.e.

$$e = -\frac{d\Phi}{dt}$$

Therefore the e.m.f. is given by:

$$e = NAB\omega \sin \omega t \text{ volts}$$

The point to note is that the e.m.f. in each turn is sinusoidal and so can be represented as shown in Figure 2.2. Also its angular frequency ($\omega = 2\pi f$) is identical to the angular frequency of rotation of the coil (f is the frequency in hertz).

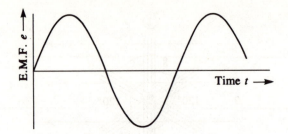

Figure 2.2 E.M.F. from a simple single-phase generator

EXAMPLE 2.1

A coil of area $10\,\text{cm}^2$ having 400 turns is rotated at 3000 rev/min in a uniform magnetic field of flux density 0.2 T. Find the peak value and hence the r.m.s. value of the voltage induced in the coil.

SOLUTION

We use the equation $e = NAB\omega\sin\omega t$. Thus the peak value of the e.m.f., $e_0 = NAB\omega$.

In this case $\omega = 2\pi \times 3000/60 = 100\pi$ radians per second, $N = 400$, $A = 10\,\text{cm}^2 = 10 \times 10^{-4}\,\text{m}^2$ and $B = 0.2\,\text{T}$. Therefore

$$e_0 = 400 \times 10 \times 10^{-4} \times 0.2 \times 100\pi = 25.13\,\text{V}$$

This is the peak value. To find the r.m.s. value we divide by $\sqrt{2}$ because the e.m.f. is sinusoidal.

Thus the r.m.s. value of e is $25.13/\sqrt{2} = 17.77\,\text{V}$.

Tutorial question 2.2

A coil of area $0.1\,\text{m}^2$ having 60 turns is rotated in a uniform magnetic field of flux density 0.8 T, to generate 50 Hz. What is the peak voltage generated?

(1508 V)

Suppose instead of just one coil we have three identical coils fixed together so that each makes an angle of 120° with the other two, as shown in Figure 2.3. We now have a simple three-phase generator, or rather, three coupled single-phase generators. Taken individually the e.m.f. in each coil

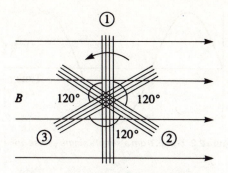

Figure 2.3 *Simple three-phase generator*

will be as shown in Figure 2.2, i.e. just the same as before, but note that coil 2 must rotate through 120° to come to the position at present occupied by coil 1, and likewise coil 3 must rotate through 240°. This is another way of saying that the e.m.f. in coil 2 is later in phase, or is lagging, by 120° with respect to coil 1, and likewise the e.m.f. in coil 3 is lagging by 240° with respect to coil 1. This is illustrated by Figure 2.4 which shows the time dependence of all three e.m.f.s. The three e.m.f.s can be expressed also as

$$e_1 = e_0 \sin \omega t$$
$$e_2 = e_0 \sin(\omega t - 120°)$$
$$e_3 = e_0 \sin(\omega t - 240°)$$
$$\text{or} \quad e_3 = e_0 \sin(\omega t + 120°)$$

where $e_0 = NAB\omega$. Note that although the angles in the brackets are expressed in degrees, they must be converted to radians ($180° = \pi$ radians) if the expressions are used for numerical calculations. The last equation

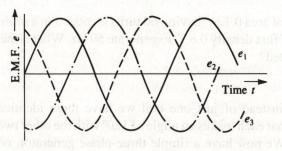

Figure 2.4 *E.M.F.s from a simple three-phase generator*

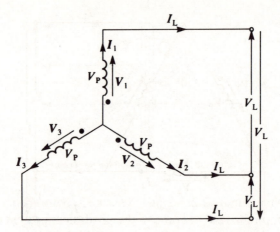

Figure 2.5 *Star connection of a three-phase generator*

shows that we can regard the e.m.f. in coil 3 as leading coil 1 by 120° rather than lagging coil 1 by 240° which, of course, is exactly the same thing. To develop the three coupled single-phase generators of Figure 2.3 into a three-phase generator, the ends of the coils must be connected together. There are two ways of doing this. The first, known as star connection, is arrived at by connecting together similar ends of the three coils to form what is called the star point. The three remaining ends are then brought out to form the three terminals of the three-phase generator. A conductor from the star point can also be brought out but this is not essential. By 'similar ends' is meant, for example, the ends of the coils at which the winding finished. This arrangement is shown in Figure 2.5 where the dots show the ends of the coils at which the winding finished. The second arrangement, known as delta connection, is shown in Figure 2.6. Here the coils form a loop, or delta, and provide just three terminals. There is no star point with a delta connection. The end of one coil is connected to the start of the next coil to form the loop. In Figures 2.5 and 2.6, V_P represents the magnitude of the voltage induced in each coil, while V_1, V_2 and V_3 are the phasors representing the voltages induced in coils 1, 2 and 3 respectively. Note that in each case the voltage phasors V_1, V_2 and V_3 are induced in the same direction with respect to the dots.

At this stage we must distinguish between phase and line quantities. The voltage across the ends of each coil is known as the phase voltage and its root mean square (r.m.s.) value denoted by V_p. Clearly $V_P = e_0/\sqrt{2}$ since

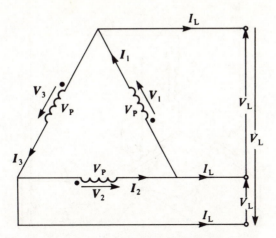

Figure 2.6 *Delta connection of a three-phase generator*

the voltage is sinusoidal. From the symmetry of the arrangement of the coils the r m.s. voltages between any two of the three terminals have the same magnitude. This is known as the line voltage V_L. Figure 2.6 shows that for a delta connection the line voltage is equal to the phase voltage. However, V_P is clearly not equal to V_L for the star connection. In fact, $V_L = |V_1 - V_2|$ (or $|V_2 - V_3|$ or $|V_3 - V_1|$), i.e. the magnitude of $(V_1 - V_2)$, etc., and the relationship between these quantities can be seen more clearly from Figure 2.7. In this diagram the lengths of the arrows represent the r.m.s. values of the voltages, while the orientations of the arrows show the phase relationships between the voltages. To obtain the relationship between V_L and V_P it is perhaps easier to consider Figure 2.8 which is similar to Figure 2.7 but with V_L transposed to the right-hand side.

From the triangle containing V_L, we have, by the sine rule

$$\frac{V_L}{\sin 120°} = \frac{V_P}{\sin 30°}$$

Therefore

$$\frac{V_L}{\sqrt{3}/2} = \frac{V_P}{1/2}$$

Therefore

$$V_L = \sqrt{3}V_P \quad \text{(star connection)}$$

Note also that V_L leads V_1 by 30°.

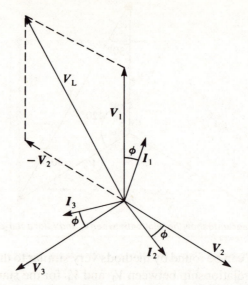

Figure 2.7 *Phasor diagram for a star-connected generator*

The relation between V_L and V_P for a star connection can also be derived from the expressions for the instantaneous phase voltages, as follows:

$$v_L = v_1 - v_2$$

Therefore

$$v_L = v_0 \sin \omega t - v_0 \sin(\omega t - 120°)$$

Therefore

$$v_L = 2v_0 \cos(\omega t - 60°) \sin 60°$$

Therefore

$$v_L = \sqrt{3} v_0 \cos(\omega t - 60°)$$

or

$$v_L = \sqrt{3} v_0 \sin(\omega t + 30°)$$

Hence the peak value of the line voltage is $\sqrt{3}$ times the peak value of the phase voltage. To obtain r.m.s. values we divide the peak value of each side by $\sqrt{2}$ and obtain

$$V_L = \sqrt{3} V_P \quad \text{(star connection)}$$

The voltage quoted by engineers working with three phase is always the line voltage, unless otherwise stated, and so the suffix L is often dropped.

Figure 2.5 shows that for a star connection the line current is equal to the phase current, i.e. $I_L = I_P$. However, Figure 2.6 clearly shows that I_L is not equal to I_P for a delta connection. The relationship between I_L and I_P for a

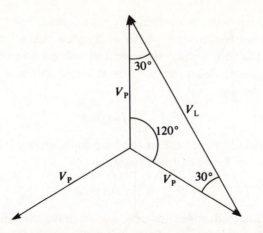

Figure 2.8 *Phasor diagram to show the relation between V_L and V_P for a star-connected generator*

delta connection can be found by methods very similar to those which were used to find the relationship between V_L and V_P for the star connection. In each phase, the phase current I_P will be lagging (or leading) the phase voltage V_P by an angle ϕ as shown in Figure 2.7. Figure 2.6 shows that $I_L = |I_1 - I_3|$ (or $|I_2 - I_1|$ or $|I_3 - I_2|$) and consideration of a phasor diagram similar to Figure 2.8, but drawn for line and phase currents, yields

$$\frac{I_L}{\sin 120°} = \frac{I_P}{\sin 30°}$$

Therefore
$$\frac{I_L}{\sqrt{3}/2} = \frac{I_P}{1/2}$$

Therefore
$$I_L = \sqrt{3}I_P \quad \text{(delta connection)}$$

The foregoing calculations assume the system is balanced, i.e. there are no significant differences between the three phases. In other words the magnitudes of the three-phase voltages are essentially equal and each is displaced by 120°, or an angle very near to 120°, from the other two. This is true most of the time in power systems except near large single-phase loads. Unbalanced systems will not be considered in this book. The advantage of a balanced system is that we need only calculate what happens in one phase to know what happens in the whole system.

There is a special expression for the power in a three-phase system, which can easily be derived as follows. Thinking of one phase in isolation the power is, or course, the same as for a single-phase arrangement.

Therefore, the power generated by one phase of a three-phase generator or absorbed by one phase of a three-phase load is $V_P I_P \cos \phi$, where ϕ is the phase angle between the voltage and current phasors. Each of the three phases generates (or absorbs) this power so the total power, P is given by

$$P = 3V_P I_P \cos \phi$$

For a star-connected generator or load we have shown that $V_L = \sqrt{3}V_P$ and $I_L = I_P$, so in terms of line quantities

$$P = \sqrt{3}V_L I_L \cos \phi$$

For a delta-connected generator or load we have shown that $V_L = V_P$ and $L_L = \sqrt{3}I_P$, so in terms of line quantities again the power is given by

$$P = \sqrt{3}V_L I_L \cos \phi$$

Since quantities quoted in three-phase circuits are assumed to be line quantities, unless otherwise stated, the expression for power in either a star or delta circuit can be written simply as

$$P = \sqrt{3}VI \cos \phi$$

If each phase of a balanced delta load has an impedance Z, then it is easy to show that each phase of the equivalent balanced star load has an impedance $Z/3$. This can be proved by using the standard delta-to-star transformation, or by finding the impedance looking into two nodes in each case.

EXAMPLE 2.2

A star-connected generator has an open-circuit line voltage of 11 kV. What is the generated e.m.f. in each phase?

SOLUTION

As the generator is open circuit, there will be no internal voltage drop. For a star connection

$$V_L = \sqrt{3}V_P$$

Therefore
$$V_P = V_L/\sqrt{3}$$

Therefore
$$V_P = 11/\sqrt{3} = 6.35 \text{ kV}$$

Tutorial question 2.3

A star-connected alternator has an open-circuit terminal voltage of 3.3 kV. What is the generated e.m.f. in each phase?

(1.9 kV)

EXAMPLE 2.3

A delta-connected three-phase load consisting of a resistor of 36 Ω and an inductor of reactance 15 Ω in each phase is connected to a 440 V three-phase supply. Calculate: (a) the equivalent star-connected load; (b) the line current; (c) the total real power consumption in the load.

SOLUTION

(a) We use the expression Z(star) = Z(delta)/3. Here Z(delta) = 36 + j15 Ω, so Z(star) = 12 + j5 Ω. Hence $R = 12\,\Omega$ and $X = 5\,\Omega$.

(b) To find the line current it is easier to consider the star load and work with just one phase to neutral.

The phase voltage is $440/\sqrt{3} = 254$ V. The current in one phase is given by Ohm's law for a.c. circuits: $I = V/Z$. Therefore

$$I = \frac{254}{12 + j5} \quad \text{or} \quad I = \frac{254}{169}(12 - j5) = 18.04 - j7.52\,\text{A}$$

and the magnitude of I is

$$\sqrt{(18.04)^2 + (7.52)^2} = 19.54\,\text{A}$$

(c) The expression above for the total power is $P = \sqrt{3}VI\cos\phi$, but here we do not know cos ϕ. It is easier to note that real power is only lost in the resistive part of the load and use the expression $P = 3I^2R$. This gives $P = 3 \times 19.54^2 \times 12 = 13.75\,\text{kW}$.

Tutorial question 2.4

A star-connected three-phase load absorbs 40 kW at a power factor of 0.875 lagging when supplied from a 6.6 kV line. Calculate: (a) the phase voltage; (b) the line current.

(3.8 kV; 4.0 A)

Tutorial question 2.5

A delta-connected three-phase load of $(80 + j60)\Omega$ per phase is connected to a 440 V three-phase supply. Calculate: (a) the phase current; (b) the line current; (c) the total real power consumed.

(4.4 A; 7.62 A; 4.65 kW)

2.3 The grid

In the year 1926 the Central Electricity Board was set up in Britain to plan the interconnection of the larger and more efficient power stations which were in operation at that time. The work was done between 1928 and 1933 using a high-voltage network known as the grid, working at 132 kV, three phase. The advantages of the grid were threefold. First, fewer generators were required as reserve. To see that this is so, consider an isolated power station supplying its own local load. Suppose four generator sets are needed to meet the maximum load. A fifth set must be provided in case one of the four unexpectedly fails. Alternatively, if the power station were linked to a similar nearby one, only one spare set out of nine would be needed, instead of one spare set out of five. Second, fewer generators were needed to run without load (spinning reserve) to take care of sudden unexpected jumps in the demand. These could be shared, by the same reasoning as above. Third, the most economical plant could be used at times of low demand. From 1926 to 1936 spare plant dropped from 70 to 26 per cent and the average cost of generation dropped to about half.

The grid was mainly responsible for these spectacular reductions. When the grid was completed it had about 5000 route kilometres of overhead lines which were operated in seven independent regions. Each region had enough generating stations to supply its own area. The one disadvantage of the grid, or indeed of any form of interconnection of generating plant, is that more current flows into a fault (short circuit), so circuit breakers able to interrupt larger currents are needed. Originally the 132 kV switchgear had a rating of 1.5 GV A, later 2.5 GV A and finally 3.5 GV A. The significance of these figures will be explained later.

In 1948 it was realized that the benefits achieved as a result of using the grid would be even greater if the whole of Britain were strongly interconnected. Thus a 275 kV network was superimposed on the 132 kV grid. The 275 kV system was built to carry large quantities of power right across the country. This is referred to as the bulk transmission of power and allows power stations to be built at the most economic sites which may be a long way from the load centre which they supply. It also allows nuclear stations to be built well away from large centres of population.

Since 1965 expansion of the grid has been made at 400 kV and some of the earlier 275 kV lines have been uprated to 400 kV. The higher voltage is more economical for large power transfers, and the addition of 400 kV lines resulted in fewer lines being required than would have been the case if the voltage had been kept down to 275 kV.

The building of the 275 kV and 400 kV system enabled the proportion of spare plant to be reduced even further.

Some of the main features of the 275 kV and 400 kV grid in England and Wales are as follows:

1. A complex network in the London area, a large industrialized urban area.
2. A complex network round Birmingham, also a large industrialized urban area.
3. A complex network in the area surrounded by the cities of Manchester, Leeds, York, Hull, Nottingham and Sheffield. This area includes the power stations built on the Yorkshire, Nottinghamshire and Derbyshire coalfields.
4. Links between the above three regions.
5. Links between South Wales and London.
6. Links between South West England and London.
7. Links from the North Wales nuclear and pumped storage stations to the Merseyside and Manchester areas.
8. Links to Scotland.

2.4 Summary

A three-phase system costs less than the equivalent single-phase system. In a star-connected generator or load $V_L = \sqrt{3}V_P$ and $I_L = I_P$. In a delta-connected generator or load $V_L = V_P$ and $I_L = \sqrt{3}I_P$. In both star and delta connections the real power $P = \sqrt{3}VI\cos\phi$. The grid was developed to interconnect groups of generators with a 132 kV network. Later the whole of Britain was interconnected by a 275 kV and 400 kV system.

CHAPTER 3

Real power, apparent power and reactive power

In this chapter we shall introduce the three kinds of power that power engineers use, and explain how and why they use them. An example shows how it is much easier to use real power balance and reactive power balance, rather than working with phase angles. This example also shows how capacitors can be used to reduce the reactive power demand of a load.

In a power system there are lines, cables, loads, etc., and each of these has its own impedance Z which can be represented as a phasor by a complex number, i.e. $Z = R + jX$ where R is the resistive component of Z, and X is the reactive component. For inductive components jX will be positive, while for capacitive components jX will be negative. Also each generator in the system which is producing power will be delivering current at some power factor $\cos \phi$ which will differ for each machine. In principle, given the voltage, power and power factor for every component of the system, it would be possible to make calculations of the power flow or current flow in every part of the system. However, calculations made this way are very difficult, mainly because there is no simple correlation between the power factors of the different components. Instead, power engineers work with reactive power, rather than power factor. This makes the calculations easier because there is a simple correlation between the reactive power in different components, but first let us define the different types of power.

The useful power, which is just called power in other branches of electrical engineering, needs an adjective to distinguish it from the other sorts of power. It is sometimes called active power, but in this book it will be referred to as real power. In a single-phase system with sinusoidal quantities it can be calculated from the product of the voltage, the current and the cosine of the phase angle between them, i.e. $VI \cos \phi$. It can also be thought of as the product of the voltage and the component of current in phase with the voltage. Real power is denoted by the symbol P and is measured in watts. Another useful quantity is the product of the voltage and current, taking no account of the phase angle. This is known as the apparent power. It is denoted by the symbol S and is measured in volt amperes. Apparent power is useful for specifying the rating of components in which the maximum voltage is fixed and the maximum current is fixed, irrespective of the phase angle. Calculations of apparent power are also useful as an intermediate step in finding the real power and the reactive power. The reactive power is the product of the voltage, the current and the *sine* of the phase angle between them, i.e. $VI \sin \phi$. It may also be thought of as the product of the voltage and the component of current in quadrature with the voltage. Reactive power is denoted by the symbol Q and is measured in volt amperes reactive (V Ar), sometimes spoken of as 'vars'. There is a tendency for power engineers to refer to these three types of power by their units rather than their names, and they speak of, for example, 'the volt amp rating', 'an increase in vars' and 'the megawatt output'.

In three-phase systems the component voltage is conventionally replaced by the line voltage which introduces a factor of $\sqrt{3}$ into the expressions for power. To summarize we have then:

Real power, $P = \sqrt{3}VI \cos \phi$ measured in watts (W).
Apparent power, $S = \sqrt{3}VI$ measured in volt amperes (V A).
Reactive power, $Q = \sqrt{3}VI \sin \phi$ measured in volt amperes reactive (V Ar).

Thus $P = S \cos \phi$ and $Q = S \sin \phi$. (A quantity met with in more advanced studies of power systems is complex power or vector power which is defined as $P+ jQ$.)

The sign of Q is arbitrary, but by convention is taken as positive at a load when the power factor is lagging. The convenience of this convention is that a load with some inductance (most loads are inductive) will consume real power and consume reactive power. A capacitor is thought of as supplying reactive power to the system, rather than consuming negative reactive power. Remember, a capacitor takes a leading power factor, i.e. the current leads the voltage.

In any one second, taking a power system as a whole, the total electrical energy generated must equal the total electrical energy consumed. This follows from the fact that electrical energy cannot be stored in the system. The energy lost or gained per second is the power so *the total of the real power generated equals the total of the real power consumed*. It can also be shown that *the total of the reactive power generated equals the total of the reactive power consumed.*

It will be shown later that this last relation makes the concept of reactive power very useful in power system calculations. We shall see shortly how to make use of these relationships, but first consider single components.

EXAMPLE 3.1
How much reactive power is consumed by a perfect inductor?

SOLUTION
Let the voltage across the inductor be V, the current through it be I and the reactance be X_L. By definition the reactive power Q is given by

$$Q = VI \sin \phi$$

Now for a perfect inductor the current lags the voltage by 90° giving $\sin \phi = 1$. By Ohm's law

$$V = IX_L$$

giving

$$Q = I^2 X_L$$

Note also that $P = 0$.

Tutorial question 3.1

An impedance of $(40 + j30)\,\Omega$ carries a current of 10 A. Calculate:
(a) the real power consumed; (b) the reactive power consumed.

$$(4\,\text{kW}; \ 3\,\text{kV Ar})$$

Tutorial question 3.2

Calculate the real power and the reactive power consumed by an impedance which has an inductance of 10 H and a series resistance of 1 kΩ, when it is connected to a 230 V, 50 Hz, single-phase supply.

$$(4.87\,\text{W}, \ 15.29\,\text{V Ar})$$

EXAMPLE 3.2
How much reactive power is generated by a perfect capacitor?

SOLUTION
Let the voltage across the capacitor be V, the current through it
be I and the reactance X_C. As before

$$Q = VI \sin \phi$$

Now for a perfect capacitor the current leads the voltage by 90°
giving $\sin \phi = 1$. By Ohm's law

$$V = IX_C$$

giving

$$Q = I^2 X_C$$

However, with a capacitor it is often more convenient to work in
terms of the voltage rather than the current. Using Ohm's law
again gives

$$Q = \frac{V^2}{X_C}$$

Note that in this example the current is leading the voltage by
90° and in the previous example it was lagging by 90°. This 180°
phase change is accounted for by regarding the reactive power in
an inductor as being consumed and the reactive power in a
capacitor as being generated.

Tutorial question 3.3

A capacitor of $2\,\mu F$ is connected to a 240 V, single-phase, 50 Hz
supply. Calculate the reactive power generated by the capacitor.

(36.2 V Ar)

Tutorial question 3.4

A capacitor is added across the terminals of the impedance in
tutorial question 3.2. Calculate the size of the capacitor required
to generate 15.29 V Ar of reactive power, and thus bring the
power factor to unity. What is the apparent power rating of this
capacitor?

($0.92\,\mu F$, 15.29 V A)

Consider now a problem solved first by using phase angles and then by using reactive power.

EXAMPLE 3.3

A three-phase 50 Hz generator supplies power to a load through a transmission line of series impedance $(50+j500)\,\Omega$ per phase. The load consumes 50 kW at 11 kV with a power factor of 0.8 lagging. Three capacitors, each $0.5\,\mu\mathrm{F}$, are connected in star across the load to improve the power factor. Measurements of the line current show it to be 2.8 A. Find the generator phase angle.

SOLUTION: METHOD A

Since the system is balanced, consider just one phase to neutral. The circuit of one phase is shown in Figure 3.1. Let the load voltage be the reference phasor. Find the real and imaginary parts of the load current phasor, I_L. For a three phase system the power is given by

$$P = \sqrt{3}VI_L \cos\phi$$

Therefore

$$I_L = \frac{P}{\sqrt{3}V\cos\phi} = \frac{50\times 10^3}{\sqrt{3}\times 11000\times 0.8} = 3.28\text{ A}$$

As a phasor $I_L = 3.28\cos\phi - j3.28\sin\phi$ (minus j because the current is lagging)

Therefore $I_L = 3.28\times 0.8 - j3.28\times 0.6$

Therefore $I_L = 2.624 - j1.968\text{ A}$

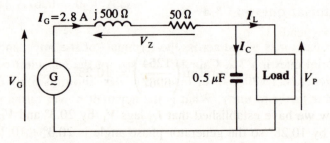

Figure 3.1 *Circuit diagram of one phase to neutral for Example 3.3*

The voltage across one phase of the load is given by

$$V_P = \frac{V}{\sqrt{3}} = \frac{11\,000}{\sqrt{3}} = 6351 \text{ V}$$

We cannot find the voltage-drop phasor along the transmission line by multiplying 2.8 by (50+j500) because we do not know the phase angle of the 2.8 A line current. Instead we must add the capacitor current to the load current.

The capacitor current is given by

$$I_C = \frac{V_P}{X_C} = j\omega C V_P = j2\pi 50 \times 0.5 \times 10^{-6} \times 6351 = j0.998 \text{ A}$$

The total current $I_L + I_C$ is also the generator current I_G so

$$I_G = I_L + I_C = 2.624 - j1.968 + j0.998$$

Therefore $$I_G = 2.624 - j0.970 \text{ A}$$

Hence the generator current lags the load voltage by

$$\arctan\left(\frac{0.970}{2.624}\right) = 20.3°$$

(Check that $|I_G| = 2.8$. $|I_G| = \sqrt{2.624^2 + 0.970^2} = 2.798$.) We still need to find the generator voltage angle. To do this we add, as phasors, the load voltage and the voltage drop along the line. The voltage drop along the line V_Z is given by

$$V_Z = I_G Z = (2.624 - j0.970) \times (50 + j500) = 616 + j1264$$

The generator voltage is

$$V_G = V_Z + V_P = 616 + j1264 + 6351 + j0 = 6967 + j1264 \text{ V}$$

so V_G leads V_L by

$$\arctan\left(\frac{1264}{6967}\right) = 10.28°$$

Now we have established that I_G lags V_L by 20.3° and V_G leads V_L by 10.28° so the generator phase angle is 20.3° + 10.28° = 30.6° lagging.

SOLUTION: METHOD B

Now see how much easier the calculation is using real and reactive power. These may be written like a profit and loss balance. The real power consumed

$$\text{in the load} = 50\,000\,\text{W}$$
$$\text{in the line} = 3 \times I^2 R = 3 \times 2.8^2 \times 50 = \quad 1\,176\,\text{W}$$

(Note the 3 here because there are three lines in a three-phase system.)

The total real power consumed in the whole system = $\overline{51\,176}$ W

This, of course, must be equal to the total real power generated, P_G.
Now find the reactive power consumed.
Since $P = \sqrt{3}VI \cos \phi$ and $Q = \sqrt{3}VI \sin \phi$ then $Q = P \tan \phi$.
In this case $\tan \phi = 0.75$.
The reactive power consumed:

$$\text{in the load} = 50\,000 \times 0.75 = 37\,500\,\text{V Ar}$$
$$\text{in the line} = 3 \times I^2 X = 3 \times 2.8^2 \times 500 = 11\,760\,\text{V Ar}$$
$$\text{Subtotal} \quad \overline{49\,260}\,\text{V Ar}$$

less the reactive power generated in the capacitors =
$3V_\text{P}^2/X_\text{C} = V^2/X_\text{C}$ (because $V = \sqrt{3}V_\text{P}$)

$$V^2/X_\text{C} = 11\,000^2 \times 2\pi50 \times 0.5 \times 10^{-6} = \underline{19\,007}\,\text{V Ar}$$
$$\text{Total} \quad \overline{30\,253}\,\text{V Ar}$$

This, of course, must be equal to the reactive power generated, Q_G.
Now from the definitions of P and Q it is easily shown that ϕ = arctan (Q/P), so the generator phase angle ϕ = arctan $(30\,253/51\,176)$ = 30.6° lagging.
This is a much easier calculation than the previous one.

Tutorial question 3.5

A star-connected load consisting of a resistor of 80 Ω and an inductor of 0.191 H in each phase is connected to a 415 V, three-phase, 50 Hz supply. Calculate: (a) the line current I; (b) the real power P consumed by the load; and (c) the reactive power Q consumed by the load. From P and Q calculate the load phase angle ϕ, and show that $P = \sqrt{3}VI \cos \phi$ and $Q = \sqrt{3}VI \sin \phi$.

(2.4 A; 1378 W; 1033 V Ar; 36.9°)

Tutorial question 3.6

Three capacitors, each $5\mu F$, are connected in delta across a 3.3 kV three-phase line. If the system frequency is 50 Hz, calculate how much reactive power is generated by the capacitors. If the system voltage falls by 5 per cent what will be the corresponding fall in reactive power generation?

(51.3 kV Ar; 9.75 per cent)

Tutorial question 3.7

Each phase of a length of 132 kV overhead power line can be represented by a series inductive reactance of 0.41 Ω and a shunt capacitive reactance to ground of 340 kΩ. At what current does the line neither generate nor consume reactive power? This condition is known as the natural loading of the line.

(204 A)

Tutorial question 3.8

A star-connected load of $(75 + j48)$ Ω per phase is supplied from a 50 Hz alternator through a transmission line of $(5+j12)$ Ω per conductor. Three capacitors, each $5\mu F$, are connected in delta across the alternator terminals. If the alternator terminal voltage is 440 V, calculate: (a) the real power; (b) the reactive power, generated by the alternator, and hence find the alternator power factor.

(1549 W; 249 V Ar; 0.987 lagging)

3.1 Summary

In three-phase systems the apparent power is $\sqrt{3}VI$, the real power is $\sqrt{3}VI\cos\phi$ and the reactive power is $\sqrt{3}VI\sin\phi$. In a system as a whole there is always a balance between the real power generated and the real power consumed. There is a similar balance for reactive power. Calculations are easier to make using real and reactive power rather than working with phase angles. Inductors consume reactive power while capacitors generate reactive power.

CHAPTER 4

Components of a power system

We start this chapter by considering how an alternator can be connected to the grid, and how it can then be represented by a simple equivalent circuit. Using this circuit we see how real and reactive power can be supplied to the grid. Next we look at some components of a power system, namely overhead lines, underground cables, switchgear and power transformers. This is followed by an explanation of the per-unit system, and the chapter ends with some symbols which enable us to draw one-line diagrams of power systems.

4.1 **The alternator**

In essence, an alternator consists of a coil which is rotated in a steady magnetic field. As explained in section 2.2 an e.m.f. is induced in the coil and the variation of the e.m.f., e, with time, t, is given by $e = NAB\omega \sin \omega t$, where A is the area of the coil, N is the number of turns on the coil, B is the flux density of the magnetic field and ω is the angular frequency of rotation of the coil. For a three-phase alternator, three symmetrically arranged coils are required. The steady magnetic field can be produced by passing a direct current through the field windings. However, large alternators are constructed the other way round, in that they have a rotating magnetic field and the generated e.m.f.s are induced in the fixed stator coils. The strength of the rotating field is controlled by a direct current fed to the rotor (the rotating part). In older generators, the rotor current is usually supplied by a machine known as an exciter, which is mounted directly on the end of the rotor shaft, and the current is fed to the rotor via brushes and slip rings. The current in the rotor is referred to as the excitation of the machine. It is much easier to feed the direct current to the rotor because, with this arrangement, the slip rings carry a much smaller current at a lower voltage than they would if the exciting field were stationary. For example, the current and

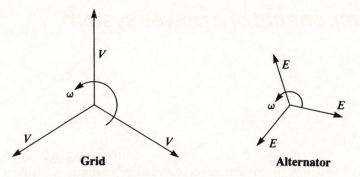

Figure 4.1 *Grid and alternator voltage phasors before synchronizing*

voltage in a typical 500 MW alternator are:

Stator, 15 000 A at 12 700 V per phase.
Rotor, 4000 A at 500 V.

Newer machines often have brushless exciters which use solid state devices to dispense with the brushes and slip rings.

Before an alternator can be connected to the grid system, four conditions must be satisfied simultaneously. The alternator must:

1. Have the same phase sequence as the grid voltages.
2. Have the same frequency as the grid voltages.
3. Have the same voltage at the circuit-breaker terminals as the grid.
4. Be instantaneously in phase with the grid.

At first only condition 1 will be satisfied; this is taken care of at the installation stage. Before the remaining conditions are satisfied, the three grid voltages and the three alternator voltages will be typically as shown in Figure 4.1 where V is the r.m.s. phase voltage of the grid at the circuit-breaker terminals, and E is the r.m.s. phase voltage of the generated e.m.f. The frequency can be increased by feeding more steam to the turbine and thus increasing the rotational speed of the turbine and alternator. The generated voltages can be increased by increasing the excitation of the alternator. The phase of the generated voltages can be matched to the phase of the grid voltages by allowing a very small frequency difference to persist. The two voltages will then very slowly drift in and out of phase in a cyclical sequence. It is then just a matter of waiting for the exact instant when they are precisely in phase, and promptly closing the circuit breakers. When the

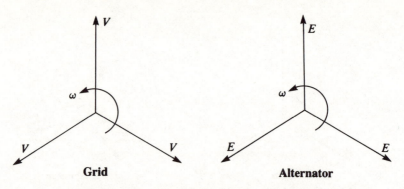

Figure 4.2 *Grid and alternator voltage phasors ready for synchronizing*

frequency, voltage and phase of the alternator have been correctly adjusted, conditions will be as shown in Figure 4.2. After the circuit breakers have been closed, the alternator is said to be synchronized to the grid. In this condition its output voltage and frequency are locked to the system values and cannot be changed by any action on the alternator, so long as it remains in synchronism with the grid. This is known as working on infinite busbars.[1]

Each phase of a synchronized alternator can be represented by the equivalent circuit shown in Figure 4.3. In many applications this can be simplified by omitting the resistive component of Z so that the equivalent circuit becomes that shown in Figure 4.4. The parameter X is known as the synchronous reactance, sometimes denoted by X_s and is constant in the normal operating situation.

Immediately after synchronizing, the situation is as shown in Figure 4.5 where the alternator is neither feeding power to, nor absorbing power from, the grid. The steam power going into the turbine which is driving the alternator is just sufficient to overcome the losses of the turbine and generator. The word 'generator' is used here to include the alternator and its exciter which provides the alternator field. If more steam is fed into the turbine one might expect the set to speed up but this is not possible when it is synchronized to the grid. Alternatively, one might expect the generator terminal voltage to rise but again this cannot happen if the generator is synchronized to the grid. What *does* happen is shown in Figure 4.6. The generator internal e.m.f., E, now leads the grid voltage by an angle δ, known as the torque angle or load angle. As a consequence current, and hence

1. An infinite busbar is an idealized concept and in practice a large machine is able to produce small voltage and frequency changes.

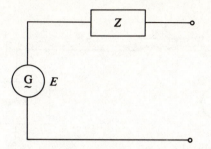

Figure 4.3 *Equivalent circuit of one phase of a synchronized alternator*

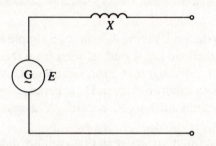

Figure 4.4 *Simplified equivalent circuit of one phase of a synchronized alternator*

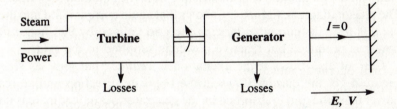

Figure 4.5 *Turbine generator synchronized but supplying no power*

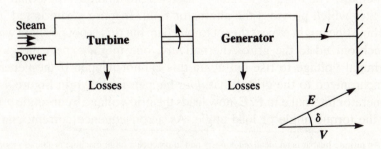

Figure 4.6 *Turbine generator supplying real power to the grid*

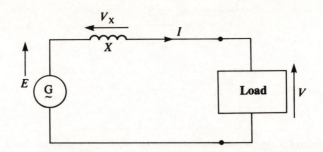

Figure 4.7 *Equivalent circuit of one phase of the generator and grid load*

power, is fed into the grid. Figure 4.7 shows a simple equivalent circuit for one phase where the load is the grid, as seen from the generator terminals. Figure 4.8 is the phasor diagram for this circuit and it shows that I is almost in phase with V, for a small angle δ, and hence real power is fed into the grid. Note that I is perpendicular to V_x because X is assumed to be a perfect inductor.

Reverting to the situation shown in Figure 4.5, what happens if the steam flow is unchanged but the alternator excitation is increased? This must increase the internal e.m.f. because an increase in excitation causes an increase in the magnetic field flux density, B, in the equation $e = NAB\omega \sin \omega t$. The phasor diagram, Figure 4.9, shows that again current is fed into the grid, but this time the current is lagging the grid voltage by 90°; hence no real power is produced but reactive power is supplied to the grid.

Current also flows if the excitation is *reduced* but now, as shown in Figure 4.10, the current flows in the opposite direction and so reactive power from the grid is absorbed by the generator. As before there is no flow of real power.

A more normal mode of operation of a turbine generator is somewhere

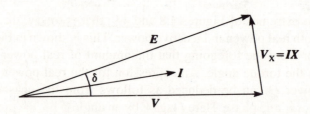

Figure 4.8 *Phasor diagram showing how real power is fed to the grid*

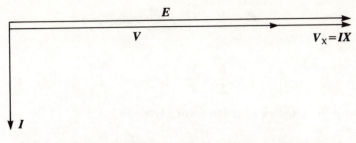

Figure 4.9 *Phasor diagram showing how reactive power is generated*

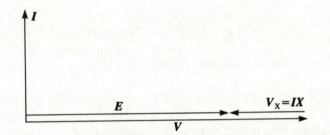

Figure 4.10 *Phasor diagram showing how reactive power is absorbed*

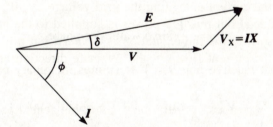

Figure 4.11 *Phasor diagram showing how real and reactive power are absorbed*

between the extremes of Figures 4.8 and 4.9; that is to say, the generator produces both real power and reactive power. This is shown in Figure 4.11.

It is clear from the foregoing that the amount of real power produced depends on the torque angle. An expression for the real power P and the reactive power Q can be deduced as follows from the phasor diagram Figure 4.11, for one phase. Here I lags V by an angle ϕ i.e. the power factor is $\cos \phi$. Thus

$$E = V + IX \quad \text{(phasors)}$$

or

$$I = \frac{E}{X} - \frac{V}{X} \quad \text{(phasors)}$$

Expressing the phasors in polar form gives

$$I = \frac{E \angle \delta}{X \angle 90°} - \frac{V \angle 0°}{X \angle 90°}$$

or

$$I = \frac{E}{X} \angle (\delta - 90°) - \frac{V}{X} \angle - 90°$$

Turning from polar to Cartesian notation we have

$$I = \frac{E}{X} \cos(\delta - 90°) + j\frac{E}{X} \sin(\delta - 90°) - \frac{V}{X} \cos(-90°) - j\frac{V}{X} \sin(-90°)$$

Now the output real power = $V \times$ (real part of I) or

$$P = \frac{VE}{X} \cos(\delta - 90°)$$

or

$$P = \frac{VE}{X} \sin \delta \quad \text{watts per phase} \tag{4.1}$$

Also the output reactive power = $V \times$ (minus imaginary part of I) or

$$Q = V\left(-\frac{E}{X} \sin(\delta - 90°) + \frac{V}{X} \sin(-90°)\right)$$

or

$$Q = \frac{VE}{X} \cos \delta - \frac{V^2}{X} \quad \text{volt amperes reactive per phase} \tag{4.2}$$

The expressions (4.1) and (4.2) derived above were obtained by considering one phase only so they should strictly speaking be written

$$P = \frac{V_P E_P}{X} \sin \delta \quad \text{watts per phase}$$

and

$$Q = \frac{V_P E_P}{X} \cos \delta - \frac{V_P^2}{X} \text{ volt amperes reactive per phase}$$

To obtain the total power we multiply each expression by 3. We can also substitute the more usual line quantities, using $V_P = V/\sqrt{3}$ and $E_P = E/\sqrt{3}$. The two $\sqrt{3}$ terms in the denominator of each expression cancel the 3 in the numerator and the expressions remain unchanged. Thus

$$P = \frac{VE}{X} \sin \delta \text{ watts}$$

and

$$Q = \frac{VE}{X} \cos \delta - \frac{V^2}{X} \text{ volt amperes reactive}$$

where E and V are line values, P is the total real power and Q is the total reactive power generated.

EXAMPLE 4.1
A 40 MVA gas turbine generator is used to absorb reactive power from a power system, i.e. act as a synchronous compensator. The terminal voltage is 11 kV and the synchronous reactance per phase is 4.5 Ω. What will be the internal e.m.f. per phase if the unit is absorbing 5 MVAr of reactive power? Assume the real power generated is negligible.

SOLUTION
We use the expression

$$Q = \frac{VE}{X} \cos \delta - \frac{V^2}{X} \tag{i}$$

We cannot solve equation (i) to find E because we do not know $\cos \delta$. We must also use the fact that the real power is zero.
Now

$$P = \frac{VE}{X} \sin \delta$$

Therefore $$\frac{VE}{X}\sin\delta = 0$$

V and E cannot be zero so δ must be zero, and $\cos \delta = 1$.
From (i)

$$Q = \frac{VE}{X} - \frac{V^2}{X}$$

Note that the reactive power is absorbed so Q is a negative quantity.

Therefore $$-5 \times 10^6 = \frac{11 \times 10^3 \times E}{4.5} - \frac{121 \times 10^6}{4.5}$$

Therefore $$E = 8.995\,\text{kV}$$

But this is the line-to-line value.
The internal e.m.f. per phase is $8.955/\sqrt{3} = 5.17\,\text{kV}$.

Tutorial question 4.1

A turbine generator supplying 290 MW to a power system has a terminal line voltage of 22 kV and an internal e.m.f. of 16 kV per phase. If the synchronous reactance of the generator is $1.25\,\Omega$ per phase, calculate the torque angle.

(36.5°)

Tutorial question 4.2

Calculate how much reactive power is generated by the machine in tutorial question 4.1.

(4.97 MV Ar)

4.2 Overhead lines

Most of the electric power which is generated in the UK is transmitted on the 275 kV and 400 kV grid. The grid conductors are suspended from insulator strings which in turn are carried on towers about 50 m high, normally spaced about 400 m apart. Each tower or pylon usually carries two three-phase circuits. Towers of three types may be distinguished: first, the straight-run towers which support only the weight of the conductors and insulators; second, the deviation towers which also resist some of the conductor tension; and third the terminal towers which are of heavier construction and are found at the end of overhead lines. These occur at substations and where lines are joined to underground cables.

The insulator strings on the grid consist of from 16 to 24 elements in series. Each element is usually a glass or porcelain disc with a metal cap on top and a metal pin on the underside. The pin fits into the cap of the next element below. Polymer insulators are also used.

Traditionally, conductors have been constructed of stranded steel-cored aluminium. That is, they consist of a few steel strands in the centre, surrounded by a larger number of aluminium strands. The steel provides the tensile strength while the aluminium provides good electrical conductivity. Aluminium was chosen because it has a higher conductivity, for a given weight, than copper and is also relatively cheap. A typical 650 A conductor has seven steel strands and 54 aluminium strands, all of which are about 3 mm in diameter. The strands are coated in grease during manufacture of the conductor to minimize corrosion of the metals.

More recently conductors with all strands of aluminium alloy have been used. These give increased capacity or lower losses and reduced maintenance.

The 132 kV distribution system conductors are used singly, but the 275 kV and 400 kV grid conductors are mounted on their insulators in groups of two or four conductors per phase, with a spacing of about 30 cm. This is known as bundling of the conductors.

There are three advantages of bundling as compared with the same volume of metal in the form of a single conductor. First, the bundle has a lower inductance. Second, there is a lower voltage gradient at the surface; this is an advantage because a high voltage gradient (electric field) at the surface of a conductor promotes corona discharges which waste power and produce radio interference. Third, bundled conductors can carry a larger current due to the improved cooling resulting from the larger surface area.

Table 4.1 shows the apparent power and current carrying capacity of various overhead lines in the UK. The figures given are for a conductor temperature restricted to 50 °C and for average conditions. In winter the ratings are about 22 per cent higher and in summer about 22 per cent

Table 4.1 *Carrying capacity per circuit of overhead lines*

Voltage (kV)	Number of conductors per bundle	Nominal area of aluminium per conductor, actual area in brackets (mm²)	Capacity (MV A)	Current (A)
132	1	175 (183)	100	450
132	1	400 (429)	150	650
275	2	175 (183)	430	900
275	2	400 (429)	620	1300
400	2	400 (429)	900	1300
400	4	400 (429)	1800	2600

The nominal areas are convenient rounded values.

lower. This is because the ambient temperature affects the cooling of the conductors.

In certain circumstances the conductor temperatures can be allowed to reach 65 °C or even 75 °C with a consequent increase in the carrying capacity of the lines.

The equivalent circuit for one phase of a short overhead line is shown in Figure 4.12, where L is the series inductance, R the series resistance, C the shunt capacitance and R_0 the leakage resistance. Values for the resistance R and the reactances of L and C at 50 Hz are given in Table 4.2 for 132 kV, 275 kV and 400 kV lines. The figures 1×175 mm², 2×400 mm² and 4×400 mm² refer to the number of conductors per bundle and the nominal cross-sectional area of the aluminium conductors. The value of R_0, the leakage resistance, is not given in the table because it is very variable. It depends on atmospheric conditions and the cleanliness of the insulators, but a typical value is 200 MΩ, so R_0 may be neglected in the equivalent circuit.

Although it is convenient to think of a line in terms of a lumped inductance, a lumped capacitance and a lumped resistance as shown in Figure 4.12, in reality these parameters are distributed evenly over the whole length of the line. For a short line the single lumped representation is

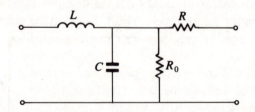

Figure 4.12 *Equivalent circuit for one phase of a short overhead line*

Table 4.2 *Resistance and reactance per kilometre length of one phase of overhead lines to British designs*

Line	132 kV	275 kV	400kV
	$1 \times 175 \text{ mm}^2$	$2 \times 400 \text{ mm}^2$	$4 \times 400 \text{ mm}^2$
R	$0.178 \,\Omega$	$0.039 \,\Omega$	$0.020 \,\Omega$
X_L	$j0.40 \,\Omega$	$j0.32 \,\Omega$	$j0.278 \,\Omega$
X_C	$-j350 \text{ k}\Omega$	$-j275 \text{ k}\Omega$	$-j245 \text{ k}\Omega$

quite satisfactory but for longer lines, say over 100 km, as used in the USA, Russia and many other countries, a more exact representation is required.

Taking the values of X_L and X_C for a 1 km length of the 132 kV line from Table 4.2, the reactive power absorbed in the series reactance at a full-load current of 450 A is 81 kV Ar, and the reactive power generated in the shunt capacitance is only 17 kV Ar, so, to a rough approximation, the capacitance can be neglected at full load, giving the equivalent circuit shown in Figure 4.13. Also, since the series resistance is small compared with the reactance, neglecting the resistance leads to an error of less than 10 per cent in the series impedance. The line may therefore be represented very simply by the circuit of Figure 4.14, where V_S is the sending-end voltage and V_R is the receiving-end voltage. Note, however, that neither Figure 4.13 nor Figure 4.14 is valid at currents much below full load. Both are, of course,

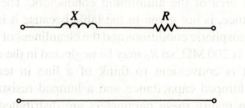

Figure 4.13 *Simplified equivalent circuit for one phase of a short overhead line*

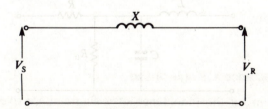

Figure 4.14 *Approximate equivalent circuit for one phase of a short overhead line*

completely wrong on no load. On no load the shunt capacitance dominates. Note also that Figure 4.14 is the same equivalent circuit as that for an alternator if V_S replaces the alternator internal e.m.f. The expression for the real power carried by the line is therefore the same as that for the real power generated by an alternator with V_S in place of E. The expression is

$$P = \frac{V_S V_R \sin \delta}{X}$$

In the case of a line, δ is known as the transmission angle.

As explained in section 4.1, the expression for P is valid in both single-phase and three-phase circuits. In a single-phase circuit V_S and V_R are the sending-end and receiving-end phase voltages, but in a three-phase circuit V_S and V_R are the sending-end and receiving-end line voltages. In each case P is the total real power carried by the line.

EXAMPLE 4.2
An overhead line 10 km long can be represented simply by a series inductive reactance of 2.7 Ω per phase. The line carries 1.8 GW with a sending-end voltage of 400 kV and a receiving-end voltage of 390 kV. What is the transmission angle?

SOLUTION
This is a short line so we can use the expression

$$P = \frac{V_S V_R}{X} \sin \delta$$

Therefore
$$\sin \delta = \frac{PX}{V_S V_R}$$

Therefore
$$\sin \delta = \frac{1.8 \times 10^9 \times 2.7}{400 \times 10^3 \times 390 \times 10^3} = 0.031\ 15$$

Therefore
$$\delta = 1.8°$$

4.2.1 Line reactance X: single phase

The expression for the inductance of a single-phase overhead line is a standard derivation (see, for example, *Electric Power Transmission and*

Distribution by P.J. Freeman). Suppose each of the two conductors has a radius r and their centres are separated by a distance d, where $d \gg r$. Then

$$L = \frac{\mu}{\pi}[\tfrac{1}{4} + \ln(d/r)] \text{ henrys per metre length}$$

In air $\mu = \mu_0 = 4\pi \times 10^{-7}$ H/m and so

$$L = 4 \times 10^{-7} \times [\tfrac{1}{4} + \ln(d/r)] \text{ henrys per metre}$$

The term $\tfrac{1}{4}$ comes from the internal flux linkage within the two conductors. For a 50 Hz system the line reactance, X_L, is therefore given by

$$X_L = 4\pi \times 10^{-5}[\tfrac{1}{4} + \ln(d/r)] \text{ ohms per metre}$$

The calculation of the line reactance for a three-phase line is difficult except in the special case of equilateral spacing.

4.2.2 Line reactance X: balanced three phase with equilateral spacing

Let the cross-section of the line be represented by Figure 4.15 where the conductors have radius r and spacing d, and $d \gg r$. It can be shown that

$$L = \frac{\mu}{2\pi}[\tfrac{1}{4} + \ln(d/r)] \text{ henrys per metre}$$

In air $\mu = \mu_0 = 4\pi \times 10^{-7}$ H/m. Therefore

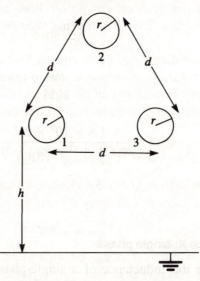

Figure 4.15 *Cross-section of a three-phase line with equilateral spacing*

$$L = 2 \times 10^{-7}[\tfrac{1}{4} + \ln(d/r)] \text{ henrys per metre per phase}$$

For a 50 Hz system the line reactance, X_L, is therefore given by

$$X_L = 2\pi \times 10^{-5}[\tfrac{1}{4} + \ln(d/r)] \text{ ohms per metre per phase}$$

Note that this is half of the value for a single-phase line and does not depend on the value of h, provided h is large, compared with d.

When the conductors of a three-phase line are not spaced equilaterally the problem of finding the inductance becomes much more difficult. The flux linkages and hence the inductances of each phase are not equal. A different inductance in each phase results in an unbalanced circuit. On a long line this can be overcome by interchanging the positions of the conductors at convenient points, e.g. substations. The average inductance per phase can be calculated by substituting d_{eq} for d in the expression above where

$$d_{eq} = (d_{12} \times d_{23} \times d_{31})^{1/3}$$

Here d_{12} is the distance between conductors 1 and 2,
$\quad\quad d_{23}$ is the distance between conductors 2 and 3
and $\quad d_{31}$ is the distance between conductors 3 and 1.

A three-phase transmission system will carry a greater power than a single-phase system, other things being equal. To show that this is so, consider a double-circuit overhead power line, i.e. a system with six independent conductors. This is the usual type of overhead line in the UK.

Suppose each conductor can carry current up to I_{max} amperes r.m.s., and the system can safely withstand a line-to-tower voltage up to V_{max} volts r.m.s. The six conductors can be used as two three-phase circuits. The maximum power that can be carried by one three-phase circuit, P, is given by

$$P = \sqrt{3}(\sqrt{3}V_{max})I_{max} = 3V_{max}I_{max}$$

because the line voltage is $\sqrt{3}$ times the line-to-tower voltage.

Two three-phase circuits can carry a maximum power P_3, given by

$$P_3 = 6V_{max}I_{max} \text{ watts}$$

Alternatively the six conductors can be used to make three single-phase circuits. The maximum power in each single-phase circuit P' is given by

$$P' = V_{max}I_{max} \text{ watts}$$

so three circuits can carry a maximum power P_1 given by

$$P_1 = 3V_{max}I_{max} \text{ watts}$$

Note that P_3 is 100 per cent greater than P_1.

Tutorial question 4.3

A 50 Hz, single-phase line consists of two parallel conductors 30 cm apart. If each conductor has a diameter of 4 mm, calculate the reactance of a 500 m length of the line. Assume the expression for the reactance per metre length.

$$(0.33 \,\Omega)$$

Tutorial question 4.4

A three-phase power line consists of three parallel conductors in the same horizontal plane. The two outer conductors are each 1 m from the centre conductor. If the conductor diameter is 6 mm, calculate the average inductance per phase of a 1 km length of the line. Assume the expression for the inductance per metre length.

$$(1.26 \,\text{mH})$$

4.3 Underground transmission cables

The 275 kV and 400 kV grid transmission lines are taken underground in urban areas and in some areas of outstanding natural beauty, e.g. Traeth Mawr in the Snowdonia National Park. The cost of taking the grid underground is so high, between 10 and 20 times the cost of an overhead line, that in most cases it is cheaper to take an overhead line on a longer route round a relatively small area of outstanding natural beauty. Cables are also less reliable by a factor of about 25. That is to say, the time for which a cable will be out of commission owing to faults will be 25 times as long as an overhead line of the same length. This is because cables need repair more often and also take longer to repair.

4.3.1 Cable construction

All insulated electric cables consist essentially of three parts: (i) the conductor; (ii) the insulation; and (iii) the external protection or sheath. The conductor is usually made of copper or aluminium, both of which have high conductivities. Many types of insulation have been tried, but traditionally high-voltage transmission cables have been insulated with oil-impregnated paper tape. Recently, cables insulated with fluid-impregnated poly-propylene/paper laminate have been developed. The insulation fluid used is a biodegradable replacement for oil.

A power system cable insulation must have the following:

1. A high dielectric strength to withstand high operating and transient voltage stresses.
2. A low power factor to minimize the heat generated.
3. A low relative permittivity to minimize the charging current.
4. Flexibility to withstand bending during installation.

The full equivalent circuit of a short length of an a.c. cable is shown in Figure 4.16 and is superficially similar to that of an overhead line. In this case, however, the shunt resistance may be thought of as two separate terms, R_0 the insulation resistance and R_d representing the dielectric loss. The other parameters are L the series inductance, R the conductor resistance and C the shunt capacitance between the conductor and the sheath. Typical values of these parameters for a 1 km length of one phase of a 400 kV grid cable with a copper conductor having an area of 2000 mm^2 are

$$L = 0.4\,\text{mH} \quad R = 9\,\text{m}\Omega \quad R_0 > 3 \times 10^{10}\Omega$$
$$R_d = 3.5\,\text{M}\Omega \quad C = 0.38\,\mu\text{F}$$

At a load current of 1.9 kA the reactive power absorbed in L is 0.45 MV Ar per km length per phase and the reactive power generated in C is 6.4 MV Ar per km length per phase. The capacitance is much more

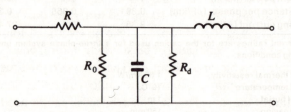

Figure 4.16 *Full equivalent circuit of a short length of cable*

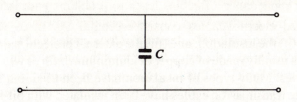

Figure 4.17 *Approximate equivalent circuit of a short cable*

significant and therefore, if the real power loss can be ignored, the cable can be represented, approximately, simply by a shunt capacitance as shown in Figure 4.17 with a typical value of $0.4\,\mu F$ per phase per km. Unlike the overhead line, this circuit is valid for any load current from zero to full load. Although it is convenient to represent the cable capacitance as a single lumped component, the capacitance is really distributed evenly over the whole length of the cable.

The current-carrying capacity of a particular cable is controlled essentially by the maximum conductor temperature at which it may be operated safely. Most transmission cables are buried underground and for these soil temperatures, soil thermal resistivity, depth of burial and, probably, the heating effect of adjacent cables will all have a bearing on their current ratings. Typical values for cables with copper conductors and corrugated aluminium sheaths, buried 1 m deep with 150 mm between phases, are given in Table 4.3.

These ratings can be increased substantially with improved cooling, for example for a $2000\,\mathrm{mm}^2$ cable in air the rating is increased to 1.8 kA and with water cooling to 2.0 kA or more.

Table 4.3 *Electrical characteristics of each phase of some high-voltage cables*

Line voltage (kV)	132	275	400
Conductor area (mm^2)	350	1000	2000
Rated current (A)	730	1090	1195
Apparent power (MV A)	170	520	825
Capacitance per phase, (μF/km)	0.391	0.368	0.377
Charging current (A/km)	9.36	18.4	27.4

The current ratings are for the cables used for a three-phase system under the following conditions:

Ground thermal resistivity	1.05 Km/W
Ground temperature	10 °C
Depth of burial	1 m
Phase spacing	150 mm
Conductor temperature	90 °C

The high capacitance of cables leads to problems with large charging currents, or, put another way, to large reactive power generation. The charging current necessitates a derating of long cables, and a critical length could be reached where the capacitance took the full rated current. For the 400 kV cable in Table 4.3 this length is 43.6 km. In practice, cables of an appreciable length are divided into sections, and reactive power compensation is applied between sections by the addition of shunt inductors. For a short cable the inductors can just be added at each end. The compensation adds significantly to the cost of an a.c. cable installation, typically adding between 20 and 25 per cent to the cost. Reactive power compensation is difficult to apply to cables under the sea and long undersea cables usually carry direct current.

Cables insulated with fluid-filled polypropylene/paper laminate (PPL) have the advantage that the dielectric loss and therefore the heat produced in the dielectric is 67 per cent less than with the traditional oil-impregnated paper tape insulation. This enables the current rating to be increased. The use of PPL also reduces the relative permittivity of the insulation by 20 per cent, which in turn reduces the capacitance by the same amount. This means that less reactive power is generated in the cable, and the cost of reactive power compensation is reduced.

4.3.2 Direct current transmission

Cables operated with d.c. offer a number of important advantages. They do not generate reactive power and so do not need reactive power compensation. They have no dielectric loss so the main loss is the I^2R loss in the conductor. Even this could be eliminated by using superconducting conductors made of, for example, niobium–tin or niobium–zirconium compounds, although at present it would not be economically viable owing to the high cost of installing and running the cooling equipment. Another advantage of a d.c. cable link is that it can be buried in a much narrower trench than an a.c. cable system, and so is more suitable for urban areas. The narrower trench width is a consequence of the much lower heat dissipation from the d.c. cables and also only two cables may be required rather than at least three for an a.c. system. There are d.c. links from Kingsnorth to Beddington and Beddington to Willesden in London.

An interesting use of d.c. cables is to interconnect two a.c. systems which are not synchronized together. The cable under the English Channel is an example which links the British system at Hythe in Kent with the French system. The cable rating is 2 GW at 270 kV d.c. The original reason for linking the two systems was to allow the transfer of electricity to France during its daily peak demand and the transfer in the reverse direction during the daily peak demand in Britain. Fortunately the peaks do not occur at the same time. In principle this could provide a useful saving of generating plant and running cost, but in practice, because of the excess of cheap nuclear generated electricity in France, the cable is used mostly to buy electricity from France.

One disadvantage of d.c. circuits is that it is very difficult, and therefore in most cases uneconomic, to provide circuit breakers to operate on high-voltage, high-current d.c. circuits. However, it is often possible to confine the switching to the a.c. part of the system. Another disadvantage of d.c. circuits is that they require rectifiers to change the a.c. to d.c. and inverters to change the d.c. back to a.c. These components add considerably to the cost of a d.c. link.

EXAMPLE 4.3

A 1 kV, single-phase, 50 Hz underground cable has a capacitance of 380 nF per km. How much reactive power is generated in a 1.5 km length of the cable?

SOLUTION

The total capacitance is $1.5 \times 380 = 570$ nF.

The expression for the reactive power Q generated in a capacitor is

$$Q = \frac{V^2}{X_C} = \omega C V^2$$

Here $\omega = 2\pi \times 50$ and $V = 1000$. Therefore

$$Q = 100\pi \times 570 \times 10^{-9} \times 1000^2 = 179 \text{ V Ar}$$

Tutorial question 4.5

A 400 kV, 50 Hz grid line is carried underground for 5 km. If the cable capacitance is 330 nF per phase per km, what is the total reactive power generated in the 5 km length?

(83 MV Ar)

Tutorial question 4.6

Calculate the charging current in each phase of the line in the previous question. If the apparent power rating of the line is 1.1 GV A, by what percentage must the maximum real power rating be reduced because of this charging current?

(120 A; 0.28 per cent)

4.4 Switchgear

In a power system switchgear is required to do the following:

1. Automatically isolate faulty components as rapidly as possible so that the remainder of the system can continue operating.
2. Allow equipment not required at a particular time to be taken out of service.
3. Isolate equipment for routine servicing or repair.
4. Prevent the spread of overvoltage surges. These are usually caused by lightning discharges on or near overhead lines.

There are three essentially different types of switchgear:

1. *Circuit breakers* which are switches designed to open, not only on full-load current, but also when carrying much higher fault currents.
2. *Fuses* which are designed to open only on fault currents. They must be replaced before the circuit can be re-energized.
3. *Isolators* which are switches that can only be opened when the circuit is dead although they can close onto a live circuit.

Modern circuit breakers are of four main types: (i) oil; (ii) air blast; (iii) sulfur hexafluoride (SF_6); and (iv) vacuum. All circuit breakers consist, in essence, of a fixed contact and a moving contact. When the circuit breaker is closed, these contacts are held together by a spring and thus allow the unimpeded flow of current. When the breaker opens, the moving contact separates from the fixed contact and draws out an arc. The arc is sustained by the flow of electrons and positive ions which are formed in the high-temperature gas. Thus it is necessary to cool the arc in order to extinguish it and this is made easier in an a.c. system by the fact that the arc current passes through zero twice each cycle. The detailed design of an oil breaker is often very complicated but the physical processes are straightforward. The heat of the arc breaks down (cracks) a small quantity of oil to form a large volume of gas, mostly hydrogen. The high pressure of this gas is used either literally to blow out the arc, or to force cold oil from a reservoir onto the arc, which is thus cooled and extinguished.

In the air-blast breaker, compressed air from a reservoir at a pressure of about 14 atmospheres is directed onto the arc at high velocities, thus extinguishing it. The air blast is usually directed along the axis of the electrodes, 'axial blast', and carries the arc through a hole in the fixed contact. Another type is the 'cross blast', in which the arc is forced onto splitter plates at the side.

Arcs can be extinguished about 20 times more efficiently in sulfur hexafluoride (SF_6) than in air. This is partly because SF_6 is a strongly electronegative gas which means that free electrons are strongly attracted to the gas molecules and thus are not free to sustain the arc. The gas is comparatively expensive and so SF_6 breakers must be equipped to recirculate the gas.

Vacuum circuit breakers are quite different in that there is no gas which can be ionized to form the arc. However, an arc can still form in ionized metal vapour which originates from the contacts. If the contacts are far enough apart when the current passes through zero, the arc will extinguish naturally. In practice the contact material has to be chosen very carefully to prevent cold welding of the contacts when they are closed and also to minimize the emission of metal vapour when they open.

The rating of a three-phase circuit breaker is conventionally expressed by multiplying the maximum current it is designed to break by the working voltage times $\sqrt{3}$, although, of course, the maximum current occurs on fault conditions when the voltage usually drops considerably. A three-phase line requires a minimum of three breaker contact pairs, one pair for each phase. The factor $\sqrt{3}$ is introduced so that the quoted rating is the apparent power for the whole system and not just for one phase. High-voltage circuit breakers often have several pairs of contacts in series in each phase. The breaking capacity of the circuit breakers on the grid has risen progressively from 1.5 GVA on the early 132 kV grid system to 35 GVA on the present 400 kV system. The rise is due partly to the increase in voltage but more especially to the increase in short-circuit current, as a consequence of more power stations being added to the system.

EXAMPLE 4.4

Each phase of a three-phase circuit breaker is capable of withstanding a voltage of 76 kV and breaking a current of 10.9 kA. What is its rating?

SOLUTION

The rating of a circuit breaker is defined as $\sqrt{3}VI$.

The breaker need only withstand the phase voltage, so the 76 kV in the question is the phase voltage. The V in the equation is the line voltage. To find the line voltage we multiply the phase voltage by $\sqrt{3}$:

Therefore
$$V = \sqrt{3} \times 76\,\text{kV}$$

and

$$\sqrt{3}VI = \sqrt{3} \times \sqrt{3} \times 76 \times 10^3 \times 10.9 \times 10^3 = 2.5\,\text{GV A}$$

Tutorial question 4.7

A circuit breaker for the 400 kV grid can break a 50.5 kA symmetrical three-phase fault current. What is the rating of the circuit breaker?

(35 GV A)

4.5 The power transformer

In general, one phase of a power transformer can be represented by a perfect transformer and four fixed components as shown in Figure 4.18, where R accounts for the winding resistance (the copper loss), R_0 accounts for the core loss, X accounts for the leakage reactance and X_0 accounts for the magnetizing current. On no load R_0 and X_0 are the dominant parameters while on full load R and X are more important. Usually X is larger than R, and if the power loss is not being calculated the transformer on load can be represented approximately by X and a perfect transformer, as shown in Figure 4.19. It will be shown in the next section that it is possible also to eliminate the perfect transformer. A perfect transformer is an idealized concept of a transformer with the following properties:

1. The primary and secondary windings have no resistance, so there is no I^2R loss in either the primary or the secondary winding.
2. The permeability of the magnetic core is infinitely large so there is no leakage flux from the core and no magnetizing current.
3. There are no losses in the core.

Thus there are no losses of any kind in a perfect transformer. If the primary

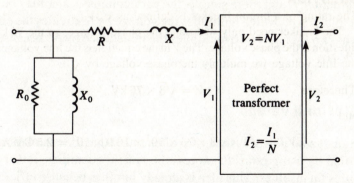

Figure 4.18 *Equivalent circuit for one phase of a power transformer*

winding has N_1 turns and the secondary winding has N_2 turns, then the transformation ratio, N, is given by

$$N = \frac{N_2}{N_1}$$

Applying an alternating voltage V_1 across the primary winding of a perfect transformer results in an alternating voltage V_2 appearing across the secondary winding where

$$V_2 = NV_1$$

If the current in the primary winding is I_1 then the current in the secondary winding I_2, is given by

$$I_2 = \frac{1}{N}I_1$$

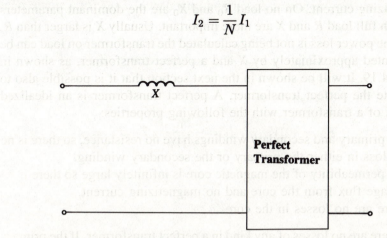

Figure 4.19 *Approximate equivalent circuit for one phase of a power transformer*

Alternatively, and more simply, the perfect transformer may be thought of as a 'black box' which multiplies the voltage by N, divides the current by N and has no losses. It also multiplies the impedance by N^2 because $V_2/I_2 = N^2 \times V_1/I_1$.

4.6 Per-unit values

When analyzing power systems it is often found that the calculations can be simplified by using ratios to express the system quantities, rather than the quantities themselves. This idea is already familiar because efficiency and losses are often expressed as percentages. A voltage V can be expressed in terms of per unit (pu) by selecting a suitable base voltage V_B and determining the ratio

$$V_{pu} = \frac{V}{V_B}$$

In general the per-unit value is

$$\frac{\text{the actual value of a quantity}}{\text{the base value of the same quantity}}$$

This equation applies to any electrical quantity, e.g. voltage, current, apparent power and impedance.

The base values of the various quantities may not be chosen at random, but must conform to the relations which normally hold between voltage, current, apparent power and impedance. Thus for a single-phase system

$$S_B = V_B I_B \quad Z_B = \frac{V_B}{I_B} \quad \text{and therefore} \quad Z_B = \frac{V_B^2}{S_B}$$

For a three-phase system

$$S_B = \sqrt{3} V_B I_B \quad I_B = \frac{S_B}{\sqrt{3} V_B} \quad Z_B = \frac{V_B}{\sqrt{3} I_B}$$

and again

$$Z_B = \frac{V_B^2}{S_B}$$

Note that in a three-phase system S_B is the total apparent power, V_B is the line-to-line voltage, I_B is the line current and Z_B is the impedance of one phase of a star load.

If the base values are chosen correctly, then the standard equations relating voltage, current, power and impedance will also be valid for per-unit quantities, e.g.

$$V_{pu} = I_{pu}Z_{pu} \quad P_{pu} = V_{pu}I_{pu}\cos\phi$$

(note that there is no $\sqrt{3}$ factor).

It is usual to select first a base value for the apparent power which will apply throughout the system. Any convenient base may be chosen. The base voltage is selected next, and the base values of current and impedance are now fixed by the above relations. Note, however, that the base voltage is not the same in all parts of a system containing transformers.

The ratio of the base voltage on the secondary side to the base voltage on the primary side of a transformer must be equal to the transformation ratio.

This is most conveniently achieved by taking as base voltages the nominal transformer voltages. If this is done, and provided all work is done in per-unit quantities, the perfect transformer may be eliminated.

It is often necessary to convert per-unit reactances from one base to another. Suppose that the base apparent power is changed from an old value S_B to a new value S_B'. If X is a reactance, then on the old base, the per-unit value of this reactance, X_{pu}, is given by

$$X_{pu} = \frac{X}{X_B}$$

or

$$X_{pu} = \frac{XS_B}{V_B^2} \tag{4.3}$$

On the new base, the per-unit value of this same reactance, X_{pu}', is given by

$$X_{pu}' = \frac{XS_B'}{V_B^2} \tag{4.4}$$

Eliminating X/V_B^2 from (4.3) and (4.4) gives

$$\frac{X_{\text{pu}}}{S_{\text{B}}} = \frac{X'_{\text{pu}}}{S'_{\text{B}}}$$

or

$$X'_{\text{pu}} = X_{\text{pu}} \frac{S'_{\text{B}}}{S_{\text{B}}} \qquad (4.5)$$

Expression (4.5) gives the rule for converting a per-unit reactance from one base apparent power to a new base apparent power.

The per-unit value of the phase voltage at a point in a system will always be the same as the per-unit value of the line voltage at that point. This means that when working with per-unit quantities we do not have to be quite so careful to distinguish phase and line voltages.

To understand why the perfect transformer can be eliminated, when working in per unit, consider the base voltages and impedances on each side of a perfect transformer. By taking base voltages proportional to the nominal transformer voltages, we ensure that the base voltage on the secondary side of each transformer is N times the base voltage on the primary side, where N is the transformation ratio. Because Z_{B} is proportional to V_{B}^2 it follows that the base impedance on the secondary side is N^2 times the base impedance on the primary side. However, the actual impedance referred to the secondary side is also N^2 times the actual impedance referred to the primary side. This means that an impedance will have the same per-unit value on each side of a transformer. In other words, transferring a per-unit impedance across a transformer does not change its value. So, provided all work is done in per-unit quantities, the perfect transformer may be eliminated.

EXAMPLE 4.5

A 20 MVA transformer with 11 kV primary and 66 kV secondary, has a reactance of 0.242 Ω referred to the primary. What is the per-unit reactance on the primary side? What is the per-unit reactance referred to the secondary?

SOLUTION

1. Let S_{B} = 20 MVA and on the primary side V_{B} = 11 kV. Then

$$Z_{\text{B}} = \frac{V_{\text{B}}^2}{S_{\text{B}}} = \frac{11^2 \times 10^6}{20 \times 10^6} = 6.05\,\Omega$$

But $X_B = Z_B$

Therefore $\qquad\qquad X_{pu} = \dfrac{X}{X_B} = \dfrac{0.242}{6.05} = 0.04\,\text{pu}$

2. Transferring $0.242\,\Omega$ to the secondary side gives $X_2 = 0.242 \times N^2\,\Omega$ where N is the transformation ratio. In this case $N = 6$.

Therefore $\qquad\qquad X_2 = 0.242 \times 6^2 = 8.712\,\Omega$

On this side S_B is still $20\,\text{MVA}$ but V_B is now $66\,\text{kV}$. Therefore

$$Z_B = \frac{V_B^2}{S_B} = \frac{66^2 \times 10^6}{20 \times 10^6} = 217.8\,\Omega$$

But $X_B = Z_B$

Therefore $\qquad\qquad X_{2pu} = \dfrac{8.712}{217.8} = 0.04\,\text{pu}$

as before.

EXAMPLE 4.6

In part of a three-phase system, the base apparent power is $100\,\text{MVA}$ and the base voltage is $132\,\text{kV}$. Calculate: (a) the base current; (b) the base impedance; (c) $210\,\text{A}$ expressed in per unit; and (d) $18\,\Omega$ expressed in per unit.

SOLUTION

(a) We use the expression for apparent power: $S = \sqrt{3}VI$ which gives $I = S/\sqrt{3}V$ and this must also hold for base values. Therefore

$$I_B = \frac{100 \times 10^6}{\sqrt{3} \times 132 \times 10^3} = 437\,\text{A}$$

(b) The base impedance is

$$Z_B = \frac{V_B^2}{S_B} = \frac{132^2 \times 10^6}{100 \times 10^6} = 174\,\Omega$$

(c) The per-unit current is

$$I_{pu} = \frac{I}{I_B} = \frac{210}{437} = 0.48\,\text{pu}$$

(d) The per-unit impedance is

$$Z_{pu} = \frac{Z}{Z_B} = \frac{18}{174} = 0.103\,pu$$

Tutorial question 4.8

A 50 Hz, 50 MV A transformer with a 132 kV primary and a 33 kV secondary has a reactance of 0.1 pu per phase. What is the reactance in ohms per star phase: (a) referred to the primary; (b) referred to the secondary?

(34.85 Ω; 2.178 Ω)

Tutorial question 4.9

A 50 Hz, 500 MV A transformer with a 400 kV primary and a 275 kV secondary has a total leakage reactance per star phase referred to the primary of 38.4 Ω. Calculate: (a) the leakage reactance referred to the secondary; (b) the per-unit reactance referred to the primary; (c) the per-unit reactance referred to the secondary.

(18.15 Ω; 0.12 pu; 0.12 pu)

Tutorial question 4.10

A 20 MV A transformer has a reactance of 0.1 pu. What is its reactance on a base apparent power of 50 MV A?

(0.25 pu)

4.7 Representation of a power system

Since, in most power systems, the three phases are balanced, it is enough in a diagram to show just one phase. It is usual also to omit the neutral line, leaving what is known as a one-line diagram. The symbols used to represent the components on a one-line diagram of a power system are shown in Figure 4.20.

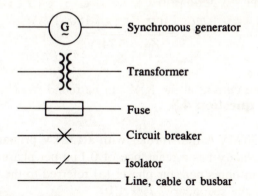

Synchronous generator

Transformer

Fuse

Circuit breaker

Isolator

Line, cable or busbar

Figure 4.20 *Symbols used to represent the components of a power system*

EXAMPLE 4.7

Figure 4.21 represents a one-line diagram of part of a power system. Find the voltage of the grid busbar, V_{sg}.

SOLUTION

First choose a base apparent power, $S_B = 100\,MV\,A$. Choose the base voltages the same as the nominal transformer voltages, i.e. 275 kV, 132 kV and 66 kV. Next find the base impedance for the line, Z_B:

$$Z_B = \frac{V_B^2}{S_B} = \frac{132^2 \times 10^6}{100 \times 10^6} = 174\,\Omega$$

Now the per-unit reactance of the line can be found from

$$X_{pu} = \frac{j3.48}{174} = j0.02\,pu$$

For the grid transformer $X_{pu} = j0.1\,pu$ (given). For the load

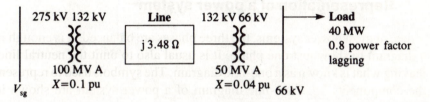

Figure 4.21 *One-line diagram of part of a power system used in Example 4.7*

transformer the per-unit reactance is given on a base of 50 MV A. This must be converted to a base of 100 MV A:

$$X_{pu} = j0.04 \times \frac{100}{50} = j0.08 \text{ pu}$$

The base current at the load can be found from

$$I_B = \frac{S_B}{\sqrt{3}V_B}$$

giving

$$I_B = \frac{100 \times 10^6}{\sqrt{3} \times 66 \times 10^3} = 875 \text{ A}$$

The actual current at the load can be found from

$$P = \sqrt{3}VI_L \cos\phi$$

giving

$$I_L = \frac{40 \times 10^6}{\sqrt{3} \times 66 \times 10^3 \times 0.8} = 437 \text{ A}$$

The per-unit value of the current I_L = 437/875 = 0.5 pu. (Alternatively note that the load current is proportional to the load MV A. The load MV A is 40/0.8 = 50 MV A and therefore $I_{pu} = S/S_B = 50/100 = 0.5$ pu.) The per-unit value of the load voltage, V_L, is 1.0 pu because the actual load voltage is the same as the base voltage.

The circuit can now be redrawn with per-unit values, as shown in Figure 4.22. The three per-unit reactances add together to give a total reactance of j0.2 pu. The circuit then reduces to the simple equivalent circuit shown in Figure 4.23.

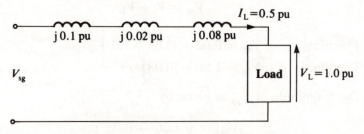

Figure 4.22 *Per-unit representation of Example 4.7*

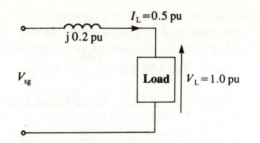

Figure 4.23 *Equivalent circuit of Example 4.7*

The problem is now equivalent to the solution of a simple single-phase circuit.

Let the phase of the load voltage, V_L, be the reference phase. The load current, I_L, can then be expressed as a phasor in the form

$$I_L = I_L \cos \phi - j I_L \sin \phi \, \text{pu}$$

(The j term is negative because the load current is lagging.) Therefore

$$I_L = 0.5 \times 0.8 - j0.8 \times 0.6 \, \text{pu}$$

$$I_L = 0.4 - j0.3 \, \text{pu}$$

The voltage drop across the inductor, V_i, is given by

$$V_i = I_L X$$

Therefore $\qquad V_i = (0.4 - j0.3) \times j0.2 \, \text{pu}$

Therefore $\qquad V_i = 0.06 + j0.08 \, \text{pu}$

The voltage of the grid busbar, V_{sg}, is given by

$$V_{sg} = V_i + V_L$$

Therefore $\qquad V_{sg} = 0.06 + j0.08 + 1.0 + j0 \, \text{pu}$

Therefore $\qquad V_{sg} = 1.06 + j0.08 \, \text{pu}$

The magnitude of V_{sg} is given by

$$|V_{sg}| = \sqrt{1.06^2 + 0.08^2} \, \text{pu}$$

Therefore $\qquad\qquad V_{sg} = 1.063\,\text{pu}$

The base voltage at the grid busbar is $275\,\text{kV}$. But $V_{pu} = V/V_B$ giving $V = V_{pu} \times V_B$. Therefore

$$V_{sg} = 1.063 \times 275\,\text{kV}$$
$$V_{sg} = 292\,\text{kV}$$

Tutorial question 4.11

Figure 4.24 represents a one-line diagram of part of a three-phase power system. Calculate the voltage of the grid busbar V_g, when the load busbar is at $11\,\text{kV}$ and the load current is $1\,\text{kA}$ at unity power factor.

$$(138.76\,\text{kV})$$

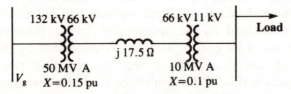

Figure 4.24 *One-line diagram of the circuit for tutorial question 4.11*

4.8 Summary

In this chapter we have looked at the major components which together form an electric power system. We have been able to represent these components by simple equivalent circuits: the alternator by an internal e.m.f. in series with a synchronous reactance; a short overhead line on load by its inductive reactance; and a cable by its capacitive reactance. We can also represent a transformer by its leakage reactance, provided we work in per-unit quantities. We have seen that per unit is a way of expressing an electrical quantity in terms of some fixed value of the same quantity, rather than expressing it as an actual value. This simplifies calculations involving transformers. The chapter concludes by showing how a three-phase system can be represented as a one-line diagram.

CHAPTER 5
Power system control

This chapter explains how frequency and voltage can be controlled in a power system, and what practical steps are taken to achieve control, first of frequency and second of voltage.

5.1 **Control of real power and frequency**

The load, or demand, on a system is continuously changing. In England and Wales the most rapid rise takes place between 6.00 a.m. and 9.00 a.m., and can be as large as 10 GW in these three hours. At most times the total mechanical output of all the turbines will not exactly balance the total electrical load on the system plus losses. However, there must always be an exact energy balance, as demanded by the law of conservation of energy. A power system has very little stored energy, but what happens is this. If the sum of the load plus losses of the system is greater than the total output of the turbines, then the extra energy comes from the rotational kinetic energy of the rotating machines which slow down. If the load plus losses of the system is less than the output of the turbines, then the rotating machines speed up. These changes of speed are reflected as changes in the frequency of the supply. On the British system a change of frequency is sensed by the turbine governors which adjust the opening of the steam valves to compensate, i.e. if the speed increases the valves start to close and if the speed decreases the valves start to open. The frequency is normally kept constant to 50 Hz ± 0.05 Hz. In practice speed control is complicated by the delay in the opening of the steam valves (0.2 ~ 0.3 s) and a further delay in the response of the intermediate and the low-pressure turbines owing to the large amount of steam trapped between the high-pressure turbines and the intermediate-pressure turbines in the reheaters (see Figure 7.1 on page 85). It is much easier to see what happens in the case of a single generator feeding its own load in isolation.

EXAMPLE 5.1

A 100 MW turbine generator is running at 3000 rev/min, 50 Hz, on no load. A 20 MW load is suddenly applied to the machine, and the governor starts to open the steam valve after 0.25 s. Calculate the frequency to which the generated voltage drops, before the steam flow increases in response to the increased load. The stored energy in the rotating parts is 400 MJ at 3000 rev/min.

SOLUTION

Before the steam valve opens the machine loses

$$20 \times 10^6 \times 0.25 = 5 \times 10^6 \, \text{J} = 5 \, \text{MJ}$$

The stored energy left is $400 - 5 = 395$ MJ. The kinetic energy of rotation is $\frac{1}{2} J \omega^2$ (J is the polar moment of inertia of the system), i.e. the kinetic energy is proportional to the frequency squared. Hence $400/395 = 50^2/f^2$ where f is the new frequency, giving

$$f = \sqrt{50^2 \times 395/400} = 49.69 \, \text{Hz}$$

Tutorial question 5.1

A turbine generator is delivering 20 MW at 50 Hz to a local load; it is not connected to the grid. The load suddenly drops to 15 MW; and the turbine governor starts to close the steam valve after a delay of 0.5 s. The stored energy in the rotating parts is 80 MJ at 3000 rev/min. What is the generated frequency at the end of the 0.5 s delay?

(50.775 Hz)

5.2 Control of reactive power and voltage

As mentioned earlier there must always be a balance between the reactive power produced and the reactive power absorbed. What happens if there is a sudden increase in the reactive power demand without a corresponding increase in the reactive power produced? Consider a simple case of a voltage source feeding a load through a line, as shown in Figure 5.1. The phasor

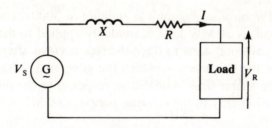

Figure 5.1 *Voltage source feeding a load through a line*

diagram corresponding to this circuit is shown in Figure 5.2 where ϕ is the phase angle of the load and δ is the transmission angle of the line. V_S and V_R are the source and load voltages respectively, of a single-phase system or equally they can be the phase voltages of a star-connected, three-phase system.

If δ is very small then the magnitude of the voltage difference between V_S and V_R, ΔV_P say, can be obtained from Figure 5.2 by resolving the phasors in a horizontal direction, which gives

$$\Delta V_P = IR \cos \phi + IX \sin \phi \qquad (5.1)$$

But $I \cos \phi = P_R/V_R$ and $I \sin \phi = Q_R/V_R$, where P_R and Q_R are the real power and the reactive power consumed by one phase of the load. Substituting in expression (5.1) gives

$$\Delta V_P = \frac{RP_R}{V_R} + \frac{XQ_R}{V_R}$$

Therefore

$$\Delta V_P = \frac{RP_R + XQ_R}{V_R} \qquad (5.2)$$

For a three-phase system we conventionally work in terms of the total real

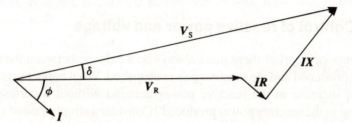

Figure 5.2 *Phasor diagram of a voltage source feeding a load through a line*

power P, the total reactive power Q, the line voltage of the load V and the line voltage drop ΔV. The conversions are

$$P = 3P_R \quad Q = 3Q_R \quad V = \sqrt{3}V_R \quad \Delta V = \sqrt{3}\Delta V_P$$

Substituting in expression (5.2) gives

$$\Delta V = \frac{RP + XQ}{V} \tag{5.3}$$

Note that X is the magnitude of the reactance and not the reactance phasor.

If $X \gg R$, as it is for the grid overhead power lines, then expression (5.3) may be simplified to

$$\Delta V = \frac{XQ}{V} \tag{5.4}$$

This means that the voltage drop depends on the reactive power flow. If there is a sudden increase in the demand for reactive power by the load, this will increase ΔV and so cause a fall in the load voltage which in turn reduces the reactive power demand.

On the British system as a whole, a reduction of voltage of 1 per cent produces a 5 per cent reduction in the reactive power demand. (It also produces a 1.4 per cent drop in the real power demand.)

Small voltage corrections can also be made by using tap-changing transformers, which enable the transformation ratio to be changed automatically by a small amount. If, say, a load voltage falls below its nominal value, by changing the tapping on the secondary, more turns can be switched into the secondary circuit of the transformer, thus increasing the secondary voltage.

EXAMPLE 5.2

A 15 km length of the 275 kV grid can be represented by a resistance of 0.51 Ω and an inductive reactance of 4.9 Ω per phase. The total load on the end of the line is 550 MW at a power factor of 0.9 lagging. Calculate the voltage drop along the line.

SOLUTION

Here $X = 4.9\,\Omega$, $R = 0.51\,\Omega$, $P = 550\,MW$ and $\cos\phi = 0.9$.
Therefore

$$\phi = 25.8°$$

Now $S = P/\cos\phi$ and $Q = S\sin\phi$. Therefore

$$S = \frac{550}{0.9} \quad \text{and} \quad Q = \frac{550}{0.9} \times \sin 25.8° = 226.4\,\text{MV Ar}$$

Using expression (5.3) derived above,

$$\Delta V = \frac{RP + XQ}{V} = \frac{0.51 \times 550 + 4.9 \times 266.4}{275}\,\text{kV} = 5.77\,\text{kV}$$

Use of the above formula for ΔV is justified only if the transmission angle, δ, is very small. It is not too difficult to calculate an exact value of ΔV by multiplying the line current phasor by the line impedance. The calculation yields a value $\Delta V = 5.92\,\text{kV}$ and shows that the simple formula is in error by only 154 V which is probably less than the error introduced in estimating the line reactance. The small error in this example is a consequence of the very small value of the transmission angle, δ, which in this case is 1.9°. Larger values of δ would, of course, lead to larger errors in using the simple equation.

EXAMPLE 5.3
A 5 km length of a 132 kV line can be represented by a resistance of 0.8 Ω and an inductive reactance of 2.05 Ω per phase. The total load on the end of the line is 80 MW at a power factor of 0.8 lagging. Calculate the voltage drop along the line: (a) neglecting the resistance; (b) including the resistance of the line.

SOLUTION
Here $X = 2.05\,\Omega$, $R = 0.8\,\Omega$, $P = 80\,\text{MW}$ and $Q = 60\,\text{MV Ar}$.
Using expression (5.4) for ΔV gives

$$\Delta V = \frac{XQ}{V} = \frac{2.05 \times 60}{132} = 0.93\,\text{kV} \quad \text{(neglecting the resistance)}$$

Using expression (5.3) for ΔV gives

$$\Delta V = \frac{RP + XQ}{V} = \frac{0.8 \times 80 + 2.05 \times 60}{132} = 1.42\,\text{kV}$$

$$\text{(including the resistance)}$$

Neglecting the resistance clearly leads to a large percentage error in ΔV. At power factors closer to unity the percentage error in ΔV is even larger.

The validity of both calculations rests on δ being very small. In this example $\delta = 0.38°$.

Tutorial question 5.2

A 10 km length of 400 kV, three-phase overhead line can be represented by an inductive reactance of 2.7 Ω per phase. The receiving-end busbars are at 400 kV when supplying a load of 1 GW at a power factor of 0.8 lagging. What is the sending-end voltage?

(405 kV)

Tutorial question 5.3

A 50 Hz, 132 kV line is 5 km long. At the receiving end there is a load of 100 MV A. The line can be represented by a resistance of 0.16 Ω per phase per km and an inductance of 1.3 mH per phase per km. Calculate the voltage drop along the line if the load power factor is: (a) 0.8 lagging: (b) 0.95 lagging, and (c) 0.93 leading.

(1.4 kV; 1.06 kV; 0.0 kV)

5.3 Generation and absorption of reactive power in a power system

Large alternators can generate reactive power up to 50 per cent of their rated real power output but only absorb reactive power up to 15 per cent of their rated output when producing full-load real power.

Shunt capacitors are used to generate reactive power but they have the disadvantage that their reactive power production is proportional to V^2. Thus, just when they are needed most, to compensate for a shortage of reactive power which has caused a fall in the local voltage, their own output falls.

Machines known as synchronous compensators can either generate or absorb reactive power. Electrically they are similar to alternators but they are not coupled to a mechanical drive. They are in fact synchronous motors with automatic voltage regulators controlling their fields. They run continuously but only consume a small amount of real power to overcome their losses. If the system voltage falls, the regulator increases the field of the machine so that it generates more reactive power. If the system voltage rises, the regulator weakens the field and the compensator absorbs reactive power. Gas turbine generators, when not needed to meet the real power demand, can often be declutched from their turbines and run as synchronous compensators.

All the components of a power system will absorb or generate reactive power. Some will do either, depending upon their mode of operation. To summarize they are listed below:

Reactive power absorption	*Reactive power generation*
Alternators underexcited	Alternators overexcited
Synchronous compensators underexcited	Synchronous compensators overexcited
Overhead lines on heavy load	Overhead lines on light load
Transformers	Underground cables
Inductors	Capacitors
A typical load	

In practice, balancing the supply with the demand, for reactive power, is made difficult by: (i) the change from generating to absorbing in overhead lines as the load is increased; (ii) the importance of keeping large quantities of reactive power from being carried by the grid; and (iii) the change in the number and location of the generating sets which are needed to meet the real power demands at different times of the day. The reason why the grid should not carry large quantities of reactive power is as follows. The rating of an overhead line is fixed by its rated voltage and maximum current. Thus the maximum apparent power, S_{max}, is fixed. If a large quantity of reactive power is carried, the power factor, $\cos \phi$, will be significantly less than unity. Thus the maximum real power which is equal to $S_{max} \cos \phi$ will be significantly reduced. This is illustrated by Example 5.4 below. Also, and probably more important, for a given real power transmitted, the losses are significantly increased owing to the increased current. This is because the largest power loss in an overhead line occurs through heating of the conductors which is proportional to $I^2 R$. This effect is illustrated by Example 5.5 below.

EXAMPLE 5.4

Find the real power that can be carried by a 400 kV quad-conductor overhead line which is carrying 1080 MV Ar of reactive power.

SOLUTION

From Table 4.1 a 400 kV quad-conductor line is rated at 1800 MV A in normal weather. Therefore

$$S = \sqrt{3}VI = 1800 \text{ MV A}$$

The reactive power, Q, is given by

$$Q = \sqrt{3}VI \sin\phi = 1080 \text{ MV Ar}$$

Therefore $\qquad \sin\phi = \dfrac{Q}{S} = \dfrac{1080}{1800} = 0.6$

Therefore $\qquad \cos\phi = 0.8$

The real power, P, is given by

$$P = S\cos\phi$$

Therefore $\qquad P = 1800 \times 0.8 = 1440 \text{ MW}$

The line will now carry only 1440 MW instead of a possible 1800 MW. This is not an economical way to use the grid.

EXAMPLE 5.5

Compare the power lost as heat in a 400 kV quad-conductor overhead line carrying 1 GW of real power only, with the same line carrying 1 GW of real power together with 750 MV Ar of reactive power.

SOLUTION

When the line is carrying 1 GW of real power only, the current, I_1, is given by

$$I_1 = \frac{P}{\sqrt{3}V\cos\phi}$$

In this case $\cos\phi = 1$. Therefore

$$I_1 = \frac{1 \times 10^9}{\sqrt{3} \times 400 \times 10^3} = 1443 \text{ A}$$

The resistance of the line is $0.020\,\Omega$ per km length (Table 4.2). The power lost in the line, P_1, is given by $P_1 = 3I_1^2R$ watts per kilometre (the 3 because there are three conductors). Therefore

$$P_1 = 3 \times 1443^2 \times 0.020 \text{ W per km}$$

Therefore $\quad P_1 = 125 \text{ kW per km}$

Now consider the line carrying 750 MV Ar of reactive power as well as the 1 GW of real power.

The current, I_2, is still given by

$$I_2 = \frac{P}{\sqrt{3}V \cos \phi}$$

But now $\phi = \arctan(750/1000)$. Therefore

$$\cos \phi = 0.8$$

Therefore $\quad I_2 = \dfrac{1 \times 10^9}{\sqrt{3} \times 400 \times 10^3 \times 0.8} = 1804 \text{ A}$

The power lost in the line, P_2, is given by $P_2 = 3I_2^2 R$ watts per kilometre. Therefore

$$P_2 = 3 \times 1804^2 \times 0.020 \text{ W per km}$$

Therefore $\quad P_2 = 195 \text{ kW per km}$

Note that P_2 is 56 per cent larger than P_1, i.e. the power lost in a line which is carrying much reactive power is significantly larger.

Tutorial question 5.4

If in tutorial question 5.2, a synchronous compensator were added to supply a current of $-j1.51 \text{ kA}$ to the receiving-end busbars, what would the sending-end voltage be? Assume the load and the receiving-end busbar voltage are unchanged.

(398 kV)

Tutorial question 5.5

A three-phase, 50 Hz line which can be represented by an inductance of 1.25 mH per phase connects the sending-end busbars to 440 V load busbars. A load taking 15 kW at a power factor of 0.8 lagging is supplied from the load busbars and three capacitors in star, each 130 μF, are also connected to the load busbars. Calculate the sending-end busbar voltage. What fall in the sending-end busbar voltage would result in the load busbars falling to 415 V? Assume the load still takes 15 kW.

(443 V; 24 V)

5.4 Summary

We have seen that if the real power demand increases, the frequency will fall unless the turbines supply more power to the system. Thus the frequency is controlled by adjusting the output power of the turbines. Also, if the reactive power demand increases, the voltage will fall unless more reactive power is generated. Thus the voltage is controlled by adjusting the reactive power supply. Reactive power can be supplied by generators, synchronous compensators and capacitors, including the natural capacitance of overhead lines and cables.

CHAPTER 6
Faults in a power system

In this chapter we shall look at the different ways in which faults can occur, and then see how to calculate the fault current in the most serious case. We shall then see how all the components of a power system can be protected from the effects of a fault.

6.1 **Types of fault**

When a component of a power system unintentionally becomes open cicuit, a fault has developed. However, in a power system context, the term 'fault' is usually reserved to denote electrical breakdown, in the form of an arc between phases of the system or from one or more phases to earth. Since faults cannot be avoided altogether, it is important to know how the system will react to a fault at any particular point.

Figure 6.1 shows the types of fault which can occur: here the three horizontal lines represent the three phases of a system. Type (i), breakdown from any one conductor to earth, is the common form of fault, but types (iii), breakdown from all three phases to earth, and (v), breakdown between all three phases, put the greatest perturbation on the system. These two types of fault are in fact equivalent, although they do not look it at first sight; also, being symmetrical, the calculation of fault current can be made in terms of one phase only. This type of fault is known as a symmetrical three-phase fault and since it is the most serious from the point of view of the stability of the system and also the easiest to calculate, we shall confine our attention to such faults.

On any system other than a very small system, the detailed calculation of the fault current is very complex and usually unnecessary. The calculations can be made much easier if a number of simplifying assumptions are made, as follows:

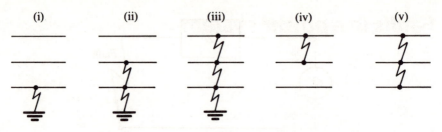

Figure 6.1 *Types of fault which can occur on a three-phase line*

1. Each generator produces its nominal e.m.f. which remains unchanged by the fault.
2. All the generators are in phase with each other.
3. The reactance of each generator drops to about one-fifth of its steady-state value and stays constant during the fault. This reduced reactance is known as the transient reactance, is denoted by X' and is often given in fault calculation problems.
4. The load currents are ignored because, in the worst case, they are much smaller than the fault current.
5. The fault forms a short circuit of zero impedance.

It is conventional to quote fault apparent power in terms of the fault current and the normal system line voltage at the point of the fault. This follows the convention used for circuit-breaker ratings, but of course assumption 5 above implies that the voltage at the point of the fault is zero. The fault apparent power is also known as the fault level.

Fault calculations are best worked in terms of per-unit quantities.

6.2 Calculation of fault current and fault apparent power

Considering only one phase to neutral and looking into the system from the point of the fault, the Thévenin equivalent of the system is a voltage source in series with an impedance as shown in Figure 6.2. Here V_T is the Thévenin equivalent *phase* voltage. Clearly the short-circuit current, I_{sc}, is given by

$$I_{sc} = \frac{V_T}{Z} \qquad (6.1)$$

It is more convenient to work in terms of per-unit quantities. Choose the nominal line voltages in the system as the base voltages. Choose any

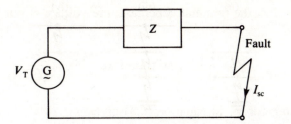

Figure 6.2 *Thévenin equivalent of one phase of a system with a fault*

convenient base apparent power S_B. Then by definition

$$S_B = \sqrt{3}V_B I_B \tag{6.2}$$

and

$$Z_{pu} = \frac{Z}{Z_B}$$

Therefore

$$Z_{pu} = \frac{ZS_B}{V_B^2}$$

because $Z_B = V_B^2/S_B$ (see section 4.6). Therefore

$$Z = \frac{Z_{pu}V_B^2}{S_B}$$

Substituting this expression for Z in equation (6.1) gives

$$I_{sc} = \frac{V_T S_B}{Z_{pu}V_B^2}$$

Substituting for S_B from equation (6.2) gives

$$I_{sc} = \frac{V_T}{Z_{pu}V_B^2} \times \sqrt{3}V_B I_B$$

Therefore

$$I_{sc} = \frac{\sqrt{3}V_T I_B}{Z_{pu}V_B} \tag{6.3}$$

Remember, we have assumed that the load currents can be neglected, which is another way of saying there are no paths to earth in the system. With no paths to earth, the Thévenin equivalent phase voltage, V_T, is equal to the nominal generated phase voltage, V_P, referred to the point of the fault, i.e. $V_T = V_P$. But the nominal generated line voltage is our base voltage so

$$V_P = \frac{V_B}{\sqrt{3}} \quad \text{and therefore} \quad V_T = \frac{V_B}{\sqrt{3}}$$

Substituting for V_T in equation (6.3) gives

$$I_{sc} = \frac{I_B}{Z_{pu}} \tag{6.4}$$

The definition of short-circuit apparent power is $\sqrt{3}$ times the nominal line voltage at the point of the fault multiplied by the short-circuit current.

To derive an expression for the short-circuit apparent power, S_{sc}, multiply each side of Equation (6.4) by $\sqrt{3}V_B$ giving

$$\sqrt{3}V_B I_{sc} = \frac{\sqrt{3}V_B I_B}{Z_{pu}}$$

The left-hand side is, by definition, S_{sc} and the numerator on the right-hand side is S_B. Therefore

$$S_{sc} = \frac{S_B}{Z_{pu}}$$

If the resistances in the system can be neglected this expression becomes

$$S_{sc} = \frac{S_B}{X_{pu}}$$

and expression (6.4) becomes

$$I_{sc} = \frac{I_B}{X_{pu}}$$

If S_{sc} and I_{sc} are expressed in per unit then

$$S_{sc\,pu} = \frac{1}{X_{pu}} \quad \text{and} \quad I_{sc\,pu} = \frac{1}{X_{pu}}$$

Note that these expressions are valid only if we take the nominal line voltages in the system as our base voltages. We are not free to choose any base voltages.

EXAMPLE 6.1

In the system shown in Figure 6.3 a symmetrical three-phase short circuit occurs on the 22 kV load busbars. Find the fault apparent power and the fault current.

SOLUTION

First choose a base apparent power of 150 MV A. This is the largest of the given ratings, but any value may be chosen and the arithmetic can sometimes be simplified by choosing a value equal to the lowest common multiple of all the apparent power ratings given in the problem. The base voltages must be the nominal busbar voltages.

For the transmission line, the per-unit reactance is given by

$$X_{pu} = \frac{X}{X_B} = \frac{XS_B}{V_B^2} = \frac{33 \times 150 \times 10^6}{132\,000^2} = 0.284\,pu$$

Convert the other reactances to a 150 MV A base using the expression derived in section 4.6:

$$0.15 \times \frac{150}{25} = 0.9\,pu \quad 0.09 \times \frac{150}{30} = 0.45\,pu$$

$$0.20 \times \frac{150}{100} = 0.3\,pu \quad 0.05 \times \frac{150}{5} = 1.5\,pu$$

(Note that a per-unit value greater than unity, i.e. a percentage value greater than 100 per cent, does not imply that a mistake has

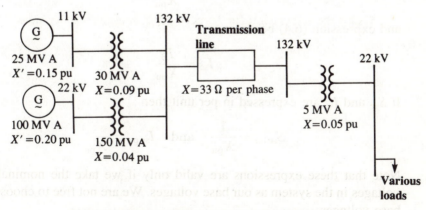

Figure 6.3 *One-line diagram of the circuit for Example 6.1*

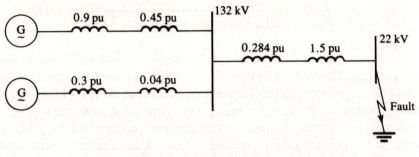

Figure 6.4 *Per-unit representation of Example 6.1*

been made. By using a suitable value of S_B any per-unit value can be obtained.)

The system can now be redrawn with all the reactances in per unit to the same base, as shown in Figure 6.4.

We can combine the two generators and their transient reactances in parallel. This is allowed because they are assumed to be in phase and generating the same per-unit voltage of unity. Now 1.35 pu reactance in parallel with 0.34 pu gives 0.272 pu. The three reactances in series give $0.272 + 0.284 + 1.5 = 2.056$ pu. The circuit becomes that shown in Figure 6.5 which is in fact the Thévenin equivalent of one phase of the system, looked at from the point of the fault. Using the expression for S_{sc} derived in section 6.2 we have

$$S_{sc} = \frac{S_B}{X_{pu}} = \frac{150 \times 10^6}{2.056} = 73\,\text{MV A}$$

Using the expression for I_{sc} derived in section 6.2 we have

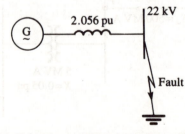

Figure 6.5 *Thévenin equivalent of one phase of Example 6.1*

$$I_{sc} = \frac{I_B}{X_{pu}} = \frac{S_B}{\sqrt{3}V_B X_{pu}} = \frac{150 \times 10^6}{\sqrt{3} \times 22\,000 \times 2.056} = 1.9\,\text{kA}$$

Sometimes fault calculations can be made without reference to the whole system. This is done by considering only that part of the system close to the fault, together with the Thévenin equivalent of the rest of the system feeding power in at some point, P, say. This point will usually be a connection to the grid. The short-circuit apparent power which the grid can supply at the point P may be quoted instead of the Thévenin equivalent circuit. This is called the short-circuit rating of the grid. It can be changed easily to a per-unit Thévenin equivalent circuit as shown in the following example.

EXAMPLE 6.2

In the system shown in Figure 6.6 there is a grid feed of 500 MV A short-circuit rating onto the 132 kV busbars at the point P. A symmetrical short circuit occurs on the 22 kV busbars. Find the fault apparent power and the fault current.

SOLUTION

The base voltage at P must be 132 kV. Choose a base apparent power of 100 MV A; this is an arbitrary choice.

The next step is to replace the grid feed by its per-unit Thévenin equivalent. To do this we use the expression $S_{sc} = S_B/X_{pu}$ which was derived earlier. Thus

$$X_{pu} = \frac{S_B}{S_{sc}} = \frac{100 \times 10^6}{500 \times 10^6} = 0.2\,\text{pu}$$

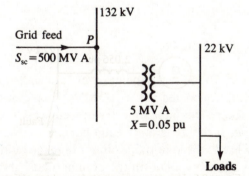

Figure 6.6 *One-line diagram of the circuit for Example 6.2*

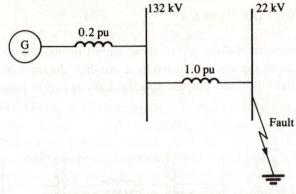

Figure 6.7 *Per-unit representation of Example 6.2*

The grid feed can therefore be thought of as a voltage source of 1.0 pu behind a reactance of 0.2 pu, on a base of 100 MV A.

Next convert the 5 MV A transformer to a 100 MV A base:

$$X_{pu} = 0.05 \times \frac{100}{5} = 1.0 \, pu$$

The diagram can now be redrawn with all the reactances in per unit to a base of 100 MV A, as shown in Figure 6.7. We can now find the short-circuit apparent power from

$$S_{sc} = \frac{S_B}{X_{pu}} = \frac{100}{1.2} \, MV A = 83 \, MV A$$

The fault current, as before, is given by

$$I_{sc} = \frac{I_B}{X_{pu}} = \frac{S_B}{\sqrt{3}V_B X_{pu}} = \frac{100 \times 10^6}{\sqrt{3} \times 22\,000 \times 1.2} = 2.2 \, kA$$

Tutorial question 6.1

Three 11 kV, 100 MV A generators are connected to common bus-bars. Each is connected via a 100 MV A inductor and an identical circuit breaker. The inductors have reactances of 0.15 pu, 0.20 pu and 0.30 pu. If the generators each have a transient reactance of 0.25 pu, what is the minimum circuit-breaker rating to protect the generators against a fault on the common busbars?

(250 MV A)

Tutorial question 6.2

A symmetrical three-phase short circuit occurs on the 22 kV busbars of the circuit shown as a one-line diagram in Figure 6.8. Calculate the fault current and the fault apparent power.

(1.9 kA, 72 MV A)

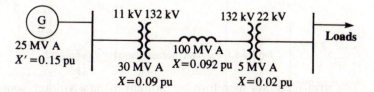

Figure 6.8 *One-line diagram of the circuit for tutorial question 6.2*

Tutorial question 6.3

A symmetrical three-phase fault occurs on the 11 kV busbars of the circuit shown in Figure 6.9. Calculate the fault apparent power and the fault current.

(353 MV A, 18.5 kA)

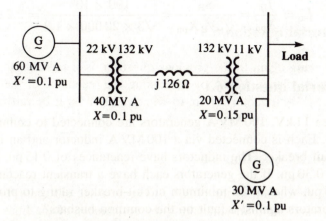

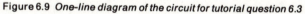

Figure 6.9 *One-line diagram of the circuit for tutorial question 6.3*

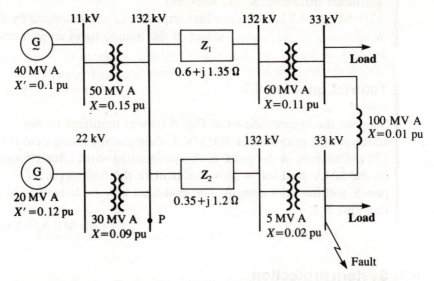

Figure 6.10 *One-line diagram of the circuit for tutorial questions 6.4, 6.5, 6.6 and 6.7*

Tutorial question 6.4

Figure 6.10 represents a one-line diagram of a power system. A symmetrical three-phase fault occurs on the 33 kV busbars as shown. Calculate the fault level, the fault current and the line voltage at the point P, under the fault condition, but neglect the resistance of each of the line impedances Z_1 and Z_2.

(210 MV A, 3.7 kA, 37 kV)

Tutorial question 6.5

Referring again to the previous question, is it possible to limit the fault current to 3.1 kA by increasing the reactance of the 100 MV A inductor? If so, to what value must it be raised?

(Yes, 0.3 pu)

Tutorial question 6.6

Referring again to the circuit in Figure 6.10, calculate the fault level and fault current, taking the resistances of the lines into account. Does the inclusion of these resistances make a

significant difference to the answers?
(210 MV A, 3.7 kA. No, the line impedances are swamped by the effect of the transformers at each end.)

Tutorial question 6.7

Suppose the system shown in Fig. 6.10 was modified by the addition of a grid feed of 500 MV A short-circuit rating onto the 132 kV busbars at the point P. A symmetrical short circuit occurs on the 33 kV busbars as shown. Calculate the fault apparent power and the fault current. The grid feed may be treated as in Example 6.2.

(384 MV A, 6.7 kA)

6.3 System protection

The generators, transformers, cables and other components of an electric power system are very expensive items and it is necessary to protect them from damage due to excessive currents and excessive voltages occurring at any point in the system. If a component develops a fault, it should be isolated from the rest of the system as rapidly as possible, to preserve the stability of the rest of the system. Another requirement is the maintenance, as far as possible, of an electrical supply to all consumers. There is one exception, however, which is mentioned on page 83. A transmission network is usually arranged so that if a fault occurs on a component, that component can be isolated by opening (tripping) circuit breakers on each side of it and so isolating the faulty component from the rest of the system. In a closely interconnected system the rest of the system can then go on working normally or at least under conditions near to normal. If the faulty component is a generator, other generators must make up the power deficiency as soon as possible. If the faulty component is an overhead line, or a transformer feeding an overhead line, then other lines will carry extra current. The system must be designed so that this extra current does not result in the other lines exceeding their maximum capacity. If alternative lines *were* overloaded, the protection system might react as if these too were faulty and isolate them from the rest of the system. This could put even more load on the remaining lines and result in these being tripped out successively. This is known as cascade tripping and is most likely to occur when the power system is heavily loaded owing to very cold weather. Cascade tripping can lead to power failures over a large area as has happened more than once in the north-east of the USA. On 9 November 1965, seven states including the city of New York and also Ontario, Canada, were affected and about 30 million people were without power, some for as long as $13\frac{1}{2}$ hours.

Because of the need to isolate faulty components as rapidly as possible, automatic fault detection and protection systems are employed. These can isolate a fault in less than 150 ms, which includes the time for the protection circuit to detect the fault as well as the time for the circuit breakers to open. The protection system on a large power system is complex, but the principle is easily understood if we consider a simple arrangement. The simplest way of protecting a power system is to divide it up into a large number of sections or zones. One zone could be a length of overhead line. Circuit breakers are installed at each end of the line and the current flowing in each phase of the line is monitored by current transformers at each end. The secondary current of a current transformer is typically 1 A under normal conditions. The primary, which is one phase of the overhead line, may simply be passed through the core on which the secondary is wound. The secondary will then be a toroidal winding and the primary effectively a single turn. With most current transformers the secondary side must always be loaded when the primary current is flowing or the transformer may be damaged by the very high secondary voltage. The secondary current in the current transformer must be a measure of the current flowing in the section of overhead line where the transformer is situated. To achieve this, the transformer must operate on a linear characteristic, not only up to full load current but also in severe fault conditions which may be 20 times full load. For this reason, these current transformers are sometimes called linear couplers. The current entering each phase of a zone is compared with the current leaving the zone in a difference circuit as shown for one phase in Figure 6.11. If the difference is large, owing to a fault in this zone of the line, the difference current trips a relay which in turn trips the circuit breakers at each end of the zone in all three phases. The section of line containing the fault is thus isolated from the rest of the system.

To illustrate this principle, consider a power station feeding a remote load via an overhead line. Figure 6.11 shows, in principle, how a zone may be protected by a unit protection arrangement. In this case the zone is a length of overhead line. The diagram shows one phase but the other two phases would be protected in the same way. At each end of the zone, there is a circuit breaker and a current transformer. Ideally the two

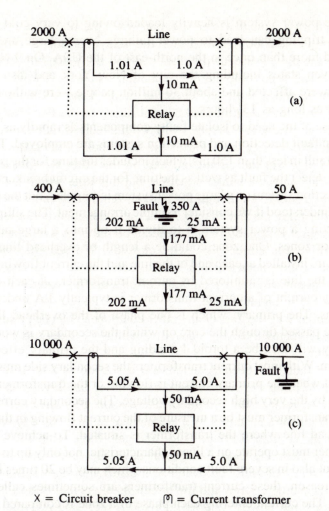

X = Circuit breaker   ^(o) = Current transformer

Figure 6.11 *Schematic diagram of an overhead line unit protection arrangement: (a) in normal operation; (b) in fault condition: and (c) carrying a through fault*

current transformers and their associated pilot wires which connect them to the relay would be identical. In practice there may be a slight imbalance. Let us assume that this imbalance is reflected as a 1 per cent difference in the pilot wire currents for the same primary current. Let us further assume that a full load current of 2000 A in the overhead line yields a current of about 1 A from each current transformer which passes along the pilot wires to the relay. Let us also assume that the relay is designed so that it requires a current greater than 100 mA to trip. Figure

6.11(a) shows a typical winter situation when the line is carrying a full-load current of 2000 A. Note that the relay current is only 10 mA, allowing for the 1 per cent imbalance so the relay does not trip. Figure 6.11(b) shows a possible summer situation when the line is carrying a light current and a fault has occurred in the zone, although the fault current is less than the full-load current rating of the line. Note that the relay current is now 177 mA; this will trip the relay which in turn will trip the circuit breakers at each end of the zone thus clearing the fault. The relay will also trip the circuit breakers in the other two phases. Figure 6.11(c) again shows a winter situation, but this time a fault has occurred in a nearby zone. Even though the fault current is five times the full-load current, the relay current is only 50 mA and the zone is not isolated. In practice the relay sensitivity is dependent on the pilot wire current, being less sensitive for larger pilot wire currents. We see that we have a protection arrangement which is sensitive to fault currents less than full-load current but will not trip on fault currents much greater than full load, if the fault is in another zone.

In a similar way the line, or any other component, can be protected from excessive voltages by using capacitive voltage dividers to monitor the phase voltages.

After a fault has been detected and a section of line isolated, it is important to bring the section back into service quickly. To do this it is necessary to reclose the circuit breakers soon after the fault has been cleared. If the fault is caused by a mechanical failure, e.g. a tower brought down by the combined effect of ice and gale-force winds, it will not be possible to clear the fault for several days. However, one of the most common causes of faults on overhead transmission lines is lightning and such faults can be cleared and the circuit breakers reclosed in less than 1 second. In some power systems high-speed reclosing relays are used to reclose the circuit breakers automatically about half a second after they have tripped. If the fault is still present they trip again. When a fault persists after a reclosure, or in some systems two reclosures, the circuit breakers remain open. In other power systems, the circuit breakers will normally be reclosed automatically after a delay of 10 to 20 s. This allows time for checks to be made on the system conditions. The checks are necessary because there are situations where reclosing the circuit breakers could result in the system becoming unstable.

An indication of the success of a closely interconnected transmission system is that less than 2 per cent of the faults on the British grid result in a loss of supply to consumers.

Each generator in a system will have its own circuit breakers which can be opened either manually or automatically if a fault develops on the generator. The generator is then said to be 'lost' from the system. Faults on generators or their associated circuits can have serious consequences, particularly when a large generator delivering full power is suddenly lost. In a small system, or a larger system which is lightly loaded, there will be a serious imbalance between the power generated and the power demand, after a large generator has been lost from the system. There may be sufficient 'spinning reserve' (see section 7.3) to cope with the imbalance but some systems use automatic load-shedding relays. This facility relies on having one or more customers to whom electricity is supplied on an interruptable basis. The supply contracts for these customers specify a maximum duration and number of interruptions per day, per week and per year. In return these customers receive their electricity at a lower cost than ordinary customers. Interruptable contracts are useful to an electricity company only if the customer normally uses a large quantity of electricity on a 24 hours per day basis. An aluminium smelter may well be supplied on this basis.

The supply is interrupted by tripping the load circuit breakers which are activated by frequency-sensitive relays. These relays continuously monitor the system frequency and if it falls below, say, 49.7 Hz, on a 50 Hz system, the first customer will be disconnected. A further fall in frequency may disconnect another customer. This gives a few minutes, during which gas turbine generators can be started or hydro-electric plant brought into operation. The system frequency will then rise and the circuit breakers will automatically reclose to restore the supply to the interrupted customers.

6.4 **Summary**

We have seen that the symmetrical three-phase fault current can be obtained from the Thévenin equivalent of one phase of the system, viewed from the point of the fault. The fault current, or short-circuit current, is calculated from $I_{sc} = I_B/Z_{pu}$ where I_B is the base current and Z_{pu} is the Thévenin impedance in per unit.

Fault protection systems consist of voltage and current sensors which are connected to relays. The relays will trip circuit breakers if a fault is detected. The circuit breakers then isolate the fault from the rest of the system, by disconnecting the faulty component.

CHAPTER 7
Energy supplies

We shall now look at the various sources of energy which can be used to generate electricity in a power system, and how the source of energy influences the design and location of each power station. Next we shall see how the demand for electricity varies during the day and with the seasons. Methods of energy storage are considered as are, finally, possible future developments in sources of energy for electricity generation.

7.1 Power stations

There are six sources of energy which together provide the primary energy for nearly all the world's electricity. They are coal (including lignite and peat), oil, nuclear energy, natural gas, water flow (hydro-electric power) and wind energy. Figure 7.1 illustrates, in schematic form, a coal or oil-burning power station.

Coal, oil and nuclear plants use the steam cycle to turn heat into electrical energy, in the following way. The steam power station uses very pure water in a closed cycle. First it is heated in the boilers to produce steam at high pressure and high temperature, typically 150 atmospheres and 550 °C in a modern station. This high-pressure steam drives the turbines which in turn drive the electric generators, to which they are directly coupled. The maximum amount of energy will be transferred from the steam to the turbines, only if the latter are allowed to exhaust at a very low pressure, ideally a vacuum. This can be achieved by condensing the outlet steam into water. The water is then pumped back into the boilers and the cycle begins again. At the condensing stage a large quantity of heat has to be extracted from the system. This heat is removed in the condenser which is a form of heat exchanger. A much larger quantity of cold, impure water enters one side of the condenser and leaves as warm water, having extracted enough heat from the exhaust steam to condense it back into water. At no point must the two water systems mix. At a coastal site the warmed impure water is

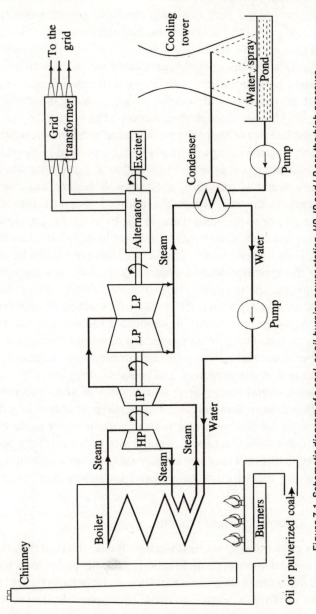

Figure 7.1 Schematic diagram of a coal- or oil-burning power station. HP, IP and LP are the high-pressure, intermediate-pressure and low-pressure turbines respectively

simply returned to the sea at a point a short distance away. A 2 GW station needs about 60 tonnes of sea water each second. This is no problem on the coast, but inland very few sites could supply so much water all the year round. The alternative is to recirculate the impure water. Cooling towers are used to cool the impure water so that it can be returned to the condensers, the same water being cycled continuously. A cooling tower is the familiar concrete structure like a very broad chimney and acts in a similar way, in that it induces a natural draught. A large volume of air is drawn in round the base, and leaves through the open top. The warm, impure water is sprayed into the interior of the tower from a large number of fine jets, and as it falls it is cooled by the rising air, finally being collected in a pond under the tower. The cooling tower is really a second heat exchanger where the heat in the impure water is passed to the atmospheric air; but unlike the first heat exchanger, the two fluids are allowed to come into contact and as a consequence some of the water is lost by evaporation. However, this is only about 1 per cent of the water which would be lost if no attempt were made to recycle the impure water. The amount lost can usually be supplied by a local river. Evaporation would eventually lead to an unacceptable build-up of impurities, and to counter this some extra water is extracted from the river while an equal quantity of the old water is returned to the river. The amount of this purge water required is about the same as that needed to make up for evaporation. The cooling towers are never able to reduce the impure water temperature right down to the ambient air temperature, so that the efficiency of the condenser and hence the efficiency of the whole station is reduced slightly compared with a coastal site. The construction of the cooling towers also increases the capital cost of building the power station. The need for cooling water is an important factor in the choice of sites for coal, oil and nuclear plants. A site which is suitable for a power station using one type of fuel is not necessarily suitable for a station using another fuel. Each may therefore be considered separately.

7.1.1 Coal-fired power stations

Early coal-burning stations were built near the load they supplied. The sites of some of the municipal stations are still to be seen today within large towns and cities. More modern stations have been built at the edge of urban areas or further away in rural situations. A station of 2 GW output, consumes about 5 million tonnes of coal in a year. In Britain where most power station coal is carried by rail, this represents an average of about 13 trains a day each carrying 1000 tonnes. This means that large coal-fired stations need a rail link unless the station is built right at the pit head. The latter has been done in a number of cases and the coal is carried directly by conveyer from the mine to the power station. It is economically attractive to build the power stations at, or close to, the coal mines because, using the grid, it is cheaper to move the electricity than the coal. This fact has led to a concentration of large coal-fired power stations over the coalfields, e.g. in the Trent Valley in England. The electricity from these stations is then moved, perhaps 100 km or more, to the load centre. This is known as the 'bulk transmission' of power. A coal-burning power station of 2 GW needs a fairly large site, because it will usually have to accommodate the following: a rail siding, preferably in the form of a loop, to avoid uncoupling the locomotive and so speed the turnround; a fuel storage area where the coal can be stockpiled during the summer months as an insurance against shortages of supply during the winter; cooling towers; a grid substation and in some cases an ash disposal area. The main power station buildings occupy a small proportion of the whole site. The site should be on, or fairly near, an existing branch of the grid if possible. Lastly, consideration has to be given to access during construction as alternator stators and grid transformers are very heavy and boiler shells are very large. Clearly, proximity to a wide main road would be an advantage.

7.1.2 Oil-fired power stations

Power station oil can be divided into crude oil, which is oil as it comes from the well, and residual oil, which remains when the more valuable fractions have been extracted in the oil refinery. The cost of moving oil by pipeline is less than that of moving coal by rail, but even so stations burning crude oil are often sited near deep-water berths suitable for unloading medium-sized tankers. Stations burning residual oil need to be sited near to the refinery which supplies them. This is because residual oil is very viscous and can only be moved through pipelines economically if it is kept warm.

7.1.3 Nuclear power stations

The heart of a nuclear power station is the reactor which takes the place of the boiler in a coal- or oil-fired power station. Most nuclear power station reactors use uranium as a fuel. Natural uranium consists of two isotopes, uranium-235 and uranium-238 ($^{235}_{92}$U and $^{238}_{92}$U) where the number 235 or 238 is the mass number, i.e. the total number of nucleons (protons and neutrons) in each nucleus, and 92 is the atomic number of uranium, i.e. the number of protons in each nucleus. Only the uranium-235 atoms can undergo fission in a thermal nuclear reactor, so only the uranium-235 constituent of the fuel elements produces the energy. Unfortunately, natural uranium contains only 0.7 per cent of uranium-235, so more than 99 per cent of the uranium which is mined and refined is waste material. In one respect it is almost like using coal with a 99 per cent ash content!

The first nuclear power stations in the UK used reactor fuel rods containing natural uranium metal in magnesium alloy cans; hence the name Magnox reactors. The nuclear chain reaction in these reactors is sustained by neutrons which have been slowed down to thermal energies by a moderator. The neutrons are slowed down inside blocks of carbon (graphite) which are placed between the fuel rods. The heat from the reactor core is extracted by a flow of carbon dioxide gas which then passes through heat exchangers to produce steam. The steam drives turbine generators in much the same way as in a coal- or oil-fired power station. These reactors are often referred to as gas-cooled reactors.

A later type known as an advanced gas-cooled reactor uses enriched uranium, i.e. uranium in which the proportion of uranium-235 has been artificially increased. The fuel is in the form of uranium dioxide (UO_2) pellets. The reason for using enriched uranium is that stainless steel cans may be used which means that advanced gas-cooled reactors can be worked at a higher temperature than Magnox reactors. This improves the steam cycle efficiency. Again carbon dioxide gas is used to extract the heat from the reactor core and graphite is used as a moderator.

For the third stage of nuclear power station development, the UK has turned to the pressurized water reactor. This reactor type was developed in the USA and other countries. It uses ordinary water under high pressure both as a moderator and to extract the heat from the reactor core. The water passes through heat exchangers so that the steam driving the turbines does not mix with the water in the reactor core. Pressurized water reactors also use enriched uranium in the form of uranium dioxide pellets. The fuel is enclosed in cans which are usually made of a zirconium alloy.

Although nuclear power stations use the steam cycle in a similar way to coal and oil plants, the site requirements for a nuclear power station are quite different. In contrast to coal, the cost of transporting nuclear fuel is negligible because of the very small amount used. A 1 GW Magnox station needs about $4\frac{1}{2}$ tonnes of uranium each week. This compares very favourably with the 50 000 tonnes of fuel which would be burnt each week

in a comparable coal-fired power station. Magnox stations use rather more cooling water than comparable coal- or oil-fired plants owing to their lower efficiency. Most nuclear power stations in the UK were built on the coast and so could use sea water for cooling. For safety reasons, early nuclear power stations were built in remote, thinly populated areas. Later sites nearer to large urban areas were chosen.

7.1.4 Natural gas power stations

Recent power station construction has moved away from large coal- and oil-burning stations. This is partly because of concern about atmospheric pollution from the sulfur in the fuel, which results in the emission of sulfur dioxide from the flue gases. This is eventually converted into sulfuric acid and falls as acid rain. Oxides of nitrogen are also produced and these are a source of more acids. It is possible to remove at least 90 per cent of the sulfur dioxide by fitting flue gas desulfurization equipment, but this adds to both the construction cost and the running cost of the power station. Large coal- or oil-burning stations are also expensive and have a long lead time, i.e. a long time between planning and completion.

Recent power station construction has favoured natural gas as a source of primary energy. Combined-cycle gas turbines achieve much higher thermal efficiencies than large coal- or oil-burning power stations. They produce virtually no sulfur dioxide, a quarter of the oxides of nitrogen and half the carbon dioxide, for the same electrical energy produced. They work in two stages, as shown in Figure 7.2. The first stage consists of one or more gas turbine generators which are essentially aircraft engines coupled to generators. The hot exhaust gases from the gas turbines are passed through a heat exchanger to produce steam for the second stage which is a conventional steam turbine generator.

An alternative way of using natural gas efficiently is again to use gas turbine generators but use the hot exhaust gases to produce steam for industrial processing or to provide space heating. This arrangement is known as a combined heat and power plant, and is also sometimes used with coal- and oil-fired plant.

7.1.5 Hydro-electric power stations

Hydro-electric power stations must be sited where the head of water is available, and as this is often in mountainous areas, they may need long transmission lines to carry the power to the nearest load centre or link up

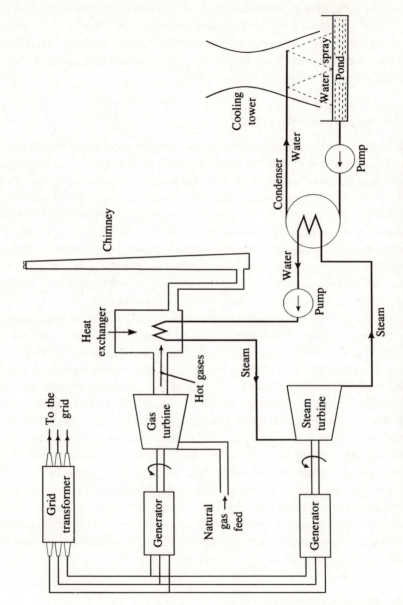

Figure 7.2 Schematic diagram of a combined-cycle gas turbine power station

with the grid. All hydro-electric schemes depend on two fundamental factors: a flow of water and a difference in level or head. The necessary head may be obtained between a lake and a nearby valley, or by building a small dam in a river which diverts the flow through the power station, or by building a high dam across a valley to create an artificial lake. With a high head of 200 m or more, considerable velocity can be imparted to the water and this can be used to drive a Pelton Wheel turbine. This is in essence a wheel with cups, or buckets as they are called, on the perimeter. The jet of water is directed into the buckets and the wheel turns in air driving the generator. The Pelton Wheel is an impulse turbine. For low heads a reaction turbine is used. This is rather like a multi-bladed propeller inside a tube through which the water flows. It is driven partly by the pressure of the water and partly by the velocity. All types of water turbines turn slowly and so if the generator, which is directly coupled to the turbine, is to generate at 50 or 60 Hz, it must have many poles. For a 50 Hz output, a typical speed is 500 rev/min, necessitating an alternator with 12 poles.

In many ways hydro-electric power stations provide the ideal way of generating electricity. They require no fuel, produce no atmospheric pollution, either chemical or thermal, produce no waste products, are cheap to run, require little maintenance, can be started up very quickly (less than 1 minute) and have a long life of at least 50 years. Unfortunately the capital costs are high. There are not many suitable sites in the UK, and most of these have already been developed. About 90 per cent of the UK's hydro-electric energy is generated in Scotland, most of it by Scottish Hydro-Electric. The largest station is Loch Sloy 160 MW. There are some small plants in Wales, e.g. Dolgarrog 34 MW and Maentwrog 30 MW in Snowdonia, and Rheidol 49 MW in Central Wales. Hydro generation represents about 2 per cent of the UK's total electricity production. Some countries are more fortunate. Norway, for example, generates almost all of its electricity requirements in hydro stations. The largest power stations in the world are hydro-electric. The Russians have built two very large power stations on the Yenisey river in Siberia: the Krasnoyarsk 6.0 GW, and the Sayano-Shushensk 6.4 GW. The largest in the USA is the Grand Coulee on the Columbia river in Washington state with a rated capacity of just under 6.2 GW. There are even larger power stations in South America. The Guri in Venezuela on the Caroni river is rated at 10.3 GW, and the largest power station in the world, the Itaipú hydro-electric plant on the Paraná river near the border of Brazil and Paraguay, has a rated capacity of 12.6 GW. It is likely that early in the 21st century, the largest hydro plants in the world will be in China.

7.1.6 Wind-powered generators

Almost all the world's primary energy sources for electricity generation originate from solar energy and nuclear energy. The solar energy is mostly obtained from sunlight which fell on the Earth millions of years ago to form coal, oil and natural gas, but recent solar energy produces rain and wind. Use of the rain to produce hydro-electric power is well established and wind energy has been used for centuries to grind corn and pump water, but it is only recently that wind energy has made a significant contribution to electricity generation.

Reserves of fossil fuels, coal, oil and natural gas, will never be completely exhausted but as the reserves are depleted the cost of extraction will rise, and a point will be reached where it is more economic to use other sources of primary energy for most electricity generation. While nuclear technology is well established, there are justifiable fears for the safety of the plants themselves and the fission products which are produced. In contrast, solar energy seems to offer an unlimited energy supply with no pollution in the form of either waste products or thermal discharges. The problems associated with using solar energy are technological and economic. At present the best way to use renewable solar energy is in the form of hydro-electric power but the next most attractive method is to make use of wind energy.

The success of a wind-powered generator depends on balancing four factors: the useful life, the capital cost, the maintenance costs and the output power. To achieve a long useful life it must be able to withstand the largest gust of wind which may occur only once in 100 years. This increases the capital cost. High-quality materials can reduce maintenance costs and extend useful life but an important factor is good design.

Very large machines will have greater output power but this may be offset by high capital cost. It may be advantageous to use a larger number of smaller machines. A large group of wind-powered generators is sometimes called a 'wind farm'. The output power from a wind turbine is proportional to the cube of the wind speed so an important factor in the economic success of a wind-powered generator is the location. It is fortunate that the most isolated places which may not be connected to the national grid are often the most windy locations and in these places wind turbines provide the most economical way of generating electricity. However, there will always be some still days when a diesel or gas turbine back-up generator will also be needed.

A typical wind turbine will have two or three blades of 25 m diameter rotating on a horizontal axis and will generate 120 kW at a wind speed of 10 m/s. Wind turbines turn quite slowly, and usually have a gearbox to increase the speed of the generator rotor.

Some very large wind farms have been assembled in the state of California, in the USA, the largest being in the Altamont Pass. By the late 1980s California had as many as 15 000 wind-powered generators with an average capacity of about 100 kW. The electricity is mostly supplied to the local grid.

7.1.7 Gas oil

Gas turbine engines, which are basically the same as the engines used to power turbo-prop aircraft, can be coupled to generators to supply electricity. Modern stations running on natural gas are described under that heading. At locations away from a natural gas main, gas turbine generators can burn gas oil which is a light fuel oil similar to paraffin. Gas oil must be highly refined to exclude impurities such as vanadium and sulfur which would otherwise shorten the life of the turbine blades. There are no special site requirements. No cooling water is required and no large fuel store, so a small site is quite adequate. Some plants have been built in urban areas, sometimes on the sites of disused coal-fired stations, where they are ideally situated for the generation of reactive power when not being used to meet the peak load demand. Other gas turbine generators are installed on the same sites as some large, modern nuclear or coal-burning stations where they take up comparatively little extra space.

Gas turbine generators are, compared with steam plant, cheap to install and also have the advantage that they can be started up in a few minutes. Their disadvantage is that they are expensive to run, when used on their own. This is due to shorter times between plant overhauls and a lower efficiency than steam plant, as well as the high fuel cost which is a consequence of the need to run on highly refined fuel. Gas oil generators are normally used in power systems for short periods to meet peak load demand.

7.1.8 Diesel oil

Large internal combustion engines similar to marine diesel engines can also be used as prime movers to generate electricity. From a power system point of view their properties are similar to simple gas turbine generators except that they run much more slowly and burn their fuel at a higher efficiency, comparable with that of a large steam plant. The fuel is, however, 35 to 40 per cent more expensive than the oil used for an oil-burning steam plant. Diesel engines are most suitable for supplementing wind generators in small isolated communities where it is not economic to make a connection to the grid.

7.1.9 Geothermal energy

The heat in the interior of the Earth can, in principle, be used to produce steam and hence drive steam turbine generators. This has only been done on a fairly small scale up to now, and in each case where the steam occurs naturally as a geyser or hot springs. The earliest plant was built at Larderello near Pisa in northern Italy, and now has an output of about 370 MW. There is a 160 MW station at Wairakei in New Zealand and several power plants at The Geysers in northern California, USA. There are also small plants in Iceland, Japan, Mexico and Russia.

7.1.10 Tidal energy

The merits of using the tidal flow of sea water to generate electrical energy have been discussed for many years. There are some places in the world where the tidal range is exceptionally large and this, combined with a suitable bay or estuary, provides a possible site for a tidal power station. Attention in the UK has focused on the Severn Estuary and in Canada on the Bay of Fundy. The world's first major tidal energy scheme is a 240 MW power station built in the Rance Estuary in St Malo Bay, France. Here the tidal range is an average of 10.9 m, and a dam was built across the mouth of the estuary so that as the tide rises, water flows through the turbines to fill the estuary and generate power. The turbines are reversible so that as the tide falls, the water flowing out of the estuary also generates power. Because of the low and variable head, the efficiency is very low, but this is not so important when there is no fuel cost. The actual source of energy is the rotational kinetic energy of the Earth combined with the gravitational pull of the Moon and the Sun.

The disadvantage of tidal schemes, from the power system point of view, is that the times of high and low tide change every day, so that sometimes the plant will be generating at times of peak load, but on other days it will be idle. It is possible, by sacrificing some of the available energy, to adjust the water levels so that the plant can generate at the time of peak demand every day. Alternatively, tidal schemes can sometimes be used so that they include a pumped storage element (see section 7.4) and in fact this is how the Rance scheme is used at present. The main objection to tidal energy plants is the high cost of the civil engineering works which are needed.

7.2 Fuel for electricity generation

The significant sources of energy for electricity generation fall into three distinct groups. They are, first, fossil fuels: coal (including lignite and peat), oil and natural gas; second, renewable energy: at present only hydro-electric and wind energy; and third, nuclear energy from uranium or plutonium. The contribution of each fuel to the total amount of fuel used by a particular power system depends primarily on the type of plant installed, and to a much lesser extent on the current fuel costs. Hydro and wind generators have zero fuel cost and nuclear stations have a low fuel cost, so the only real scope for switching fuel in response to price changes is between the fossil fuels.

By far the largest running cost item of an electricity undertaking which produces a large proportion of its energy from burning fossil fuel is the cost of the fuel itself. It is unfortunate, therefore, that this fuel can only be converted into electricity with an efficiency of between 25 and 45 per cent in many plants. Put another way, this means that more than half of the fuel bought is wasted as heat thrown away in these power stations. This heat is dissipated in the atmosphere, in a river or in the sea. In some countries, notably Denmark and Sweden, large areas of towns are heated from the cooling water of local power stations. This system is known as district heating and large, thermally insulated, water mains are laid under the streets. The condenser cooling water leaves the power station at a higher temperature than normal for a conventional power station. This results in a lower efficiency of the steam cycle, and so a lower efficiency of generation. The important factor, however, is the efficiency measured as useful output (heat and electricity) divided by the total fuel energy input. This can be as high as 80 per cent. One objection to these schemes, which is often raised, is that the homes, shops and offices do not need heating in the summer and the power station must, therefore, be provided with an alternative way of rejecting its waste heat. A more satisfactory solution is to shut down the plant during the summer. The electricity production is then lost for these months, but as the summer demand is much less than the winter demand (see the next section), other stations in the system can easily cope.

The most attractive opportunity to install a district heating scheme occurs when a new town is built in a rural area. The town and power station can then be planned together and the heating mains installed before the roads and buildings are constructed.

7.3 **Base and peak loads**

The fact that electricity cannot be stored in worthwhile quantities has a profound effect on power system management. It means, of course, that enough electricity must be generated at all times to meet the demand. A gas system has considerable storage in the mains and other pipework, but there is no storage of energy in the transmission and distribution networks of an electric power system to meet unexpected increases in the demand. The variations in demand are, therefore, very important to the power engineer. These variations can be studied on a daily, weekly and annual basis. Figure 7.3 shows the load (or demand) against time curve for the national grid system in 1993–94 for a typical winter day and a typical summer day, also the maximum winter demand and the minimum summer demand. There is a marked difference between night and day. Demand is low during the night when most people are asleep, and rises rapidly between 6.00 a.m. and 9.00 a.m. In summer this rise represents a 50 per cent increase in 3 hours, and about 10 GW of plant must be brought into operation to meet the demand. In winter, as well as the morning rise, there is an evening peak between 4.30 p.m. and 7.30 p.m. This is produced by people returning home from work and switching on lights, heaters and cookers. This evening peak presents a problem to the power engineer because it must be met by plant which is required to generate for only 2 or 3 hours each day and then only in the winter.

The average demand over a year, divided by the maximum demand which occurs during that year, is called the system load factor; typical values are 50 to 70 per cent. From the point of view of economic operation, the system load factor should be as high as possible, which explains why electricity is often cheaper during the night. This encourages the use of storage heaters which improve the system load factor by consuming more electricity during the night.

In some hot countries the summer demand is greater than the winter. This comes about where the winter is not cold enough to need much electric heating, but the summer is so hot that extensive use is made of air-conditioning plant.

An alternative way of expressing the daily load is to plot each load value

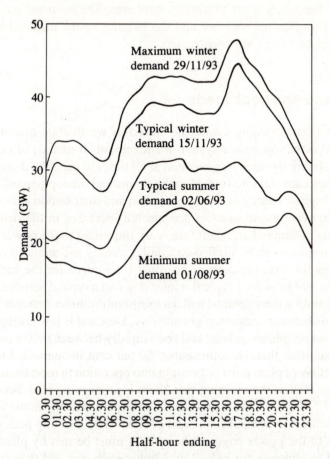

Figure 7.3 *The load (or demand) against time curve for the national grid system in 1993–4 for a typical winter day and a typical summer day, also the maximum winter demand and the minimum summer demand*

against the total length of time during each day that the particular load value is exceeded. This is known as the load–duration curve. Figure 7.4 shows the load–duration curve corresponding to a typical winter day, derived from the curve in Figure 7.3. The axes of these curves look very similar, but note that the horizontal axis of the load–time curve is 'time of day', whereas the horizontal axis of the load–duration curve is 'duration in hours'. This curve may be divided into three parts as follows:

1. The base load which must be met continuously, i.e. for 24 hours each day.

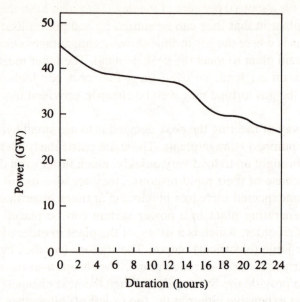

Figure 7.4 *Load–duration curve for a typical winter day derived from Figure 7.3.*

2. The intermediate load which lies between (1) and (3).
3. The peak load which is demanded for, say, 10 per cent of the time, i.e. 2 to 3 hours.

For economic reasons the base load should be supplied by the most efficient stations in terms of running cost. These will be first the nuclear and then the most modern fossil-fuel fired stations. They will be manned with a three-shift work force for 24 hours each day. In a system where there are large hydro-electric plants and plenty of water available, these may also supply some or all of the base load. Small hydro-electric plants, because of their flexibility, are better used to supply the peak demands.

The intermediate load can be met by the older, less efficient stations, and they will inevitably be idle for some part of each day. Their boilers must be kept alight and generally will produce enough steam to keep the turbine generators in synchronism with the rest of the system.

The peak load can be met in a number of ways. One way is to use gas turbine generators. These are expensive to run because of their high fuel cost and low efficiency, but this is offset by their low capital cost. If they are only used for a short time each day, their high fuel cost is not important, but if they were called upon to generate for longer times, they would be uneconomic compared with steam plant. The break-even point occurs when they are run for about 10 per cent of the day. They also have the advantage over steam plant in that they can be started up and put on load in about 2 minutes. This is where the gas turbine shows its advantages compared with using old steam plant to meet the peak demand. The latter must be run for long periods on no load, so even on a fuel cost per joule (electrical) assessment, the gas turbine may well be cheaper, provided it is not run for too long.

Another way of meeting the peak demand is to use small hydro-electric stations and pumped storage plants. These are particularly useful because they can be brought up to load very quickly, much faster even than the gas turbines. Because of their rapid response, they are also useful for coping with small, unexpected increases in demand at times other than the peak.

All the generating plant in a power system can be placed in what is known as merit order, which is a listing of the plant in order of increasing fuel cost per joule (electrical) output. The stations can then be plotted as horizontal strips on a load–duration graph. The cheapest-to-run stations are put in first to provide the base load, and then the next cheapest and so on, with the most expensive either at the top or left off altogether. Figure 7.5 shows a merit order for a hypothetical power system.

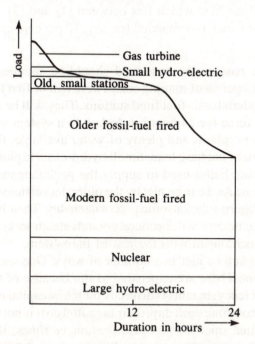

Figure 7.5 *Power station merit order for a hypothetical power system*

The large hydro stations must be at the bottom because their fuel cost is zero. Next come the nuclear stations with their low fuel cost, then the most modern fossil-fuel-fired stations with their relatively high efficiencies, followed by the older fossil-fuel-fired stations and above them the old, small stations. Above these are the small hydro plants which are clearly out of place on a strict fuel cost basis, but are placed here because of their flexibility. The top of the curve extending horizontally to a maximum of about 2 hours is supplied by the gas turbine generators. This merit order, shown in Figure 7.5, is somewhat oversimplified in that it ignores transmission costs and some other factors, but it gives the general picture of how the stations in a power system would be used. The balance between the use of coal, oil and natural gas in the fossil-fuelled power stations depends on the local costs of these fuels and can change from year to year.

Pumped storage schemes can alter the shape of the load–duration curve and will be considered in the next section.

At any particular time, some of the generators which are synchronized to the system will be running on only part load so that they can respond almost instantly (less than 1 s) to any sudden increase in demand. This is called spinning reserve by power engineers. The engineers controlling a power system try to predict a day or two in advance what the demand is likely to be at every minute of the day. To do this they study past demand records, weather forecasts and even television programme schedules. It is not the electricity consumed by the television sets themselves which is significant, but rather the habits of the people watching. At the end of a popular programme, many people stop watching and move into another room where they may turn on lights, electric fires and plug in electric kettles. Normally these actions would have a negligible effect on the power system, but if they are made almost simultaneously by perhaps a million people the effect is substantial. At such times the demand can rise by up to 2 GW within a few minutes in Britain.

The cost per joule (electrical) of supplying a small increase in the demand, at any given time, is called the incremental cost. It is equal to a weighted average of the cost of generation at those stations on part load at that particular time. These, of course, are the only stations which can supply the extra energy. At first sight the merit order would suggest that at times of low demand the incremental cost would be low. This is not necessarily true. Even at times of low demand, such as a summer Sunday afternoon, it may be necessary to start up gas turbine generators to cope with a sudden increase in demand which was not predicted, or a sudden increase which is expected to last for only a few minutes. It is true to say that the incremental cost is high at the time of peak demand, particularly in the winter when most of the plant is in use.

7.3.1 The Pool

The actual scheduling of individual generating plant is performed by the National Grid Company through an arrangement known as the pool. Each day is divided into 48 half-hour periods. On the previous day the generating companies in England and Wales are invited to bid for the supply of electricity in each of these half-hour periods. Each company will specify the quantity of electricity it can supply and the price it would charge. The National Grid Company will then choose which bids to accept, in order to schedule the required amount of generation at the lowest total cost. Generating stations in Scotland and France can also join the pool by using the interconnectors.

7.4 Energy storage

Clearly, any electric power system is not being used in the best way, from a fuel cost point of view, if efficient power stations are running at reduced power for part of the day, while generators which are expensive to run have to be used at other times. If some means were available so that energy could be stored at times of low demand, in such a way that it could be released later at times of peak demand, then there would be an overall saving of fuel. Whether there would also be an overall financial saving depends on the capital cost of constructing the storage plant. In practice, no storage system operates without losses and so the efficiency of the plant must also be taken into account. Storage will show a financial saving if the cost of the storage cycle per joule of electricity stored is less than the difference in incremental cost per joule between base load generation and peak load generation. The difference in incremental cost per joule is much larger in a system which has more than enough nuclear plant to supply the base load.

Energy can be stored in secondary batteries in the form of chemical energy, and while these are quite efficient, their high capital cost, coupled with a relatively short life, makes them uneconomic for power system use.

The hydro-electric pumped storage scheme is the only system of storing energy on a large scale at present available to the power system engineer. In its simplest form the scheme consists of a low-level reservoir and a high-level reservoir close by. When storing energy, power is drawn from the grid to drive a synchronous motor which is coupled to a hydraulic pump. This pumps water from the lower reservoir through a pipeline to the upper reservoir. In the generating mode, water flows from the upper reservoir down the pipeline and through a turbine coupled to a generator, to produce electric power in just the same way as an ordinary hydro-electric power station. It is usual to make the same machine function both as a synchronous motor and an alternator, and in some installations the pump and turbine are also the same machine which is simply reversed to change mode.

Pumped hydro-electric storage schemes require rather special geological conditions. There must be a large difference in level over as short a horizontal distance as possible, and the rock must be such that it will retain the water, particularly in the upper reservoir. They are nearly always built in mountainous regions and the lower reservoir is usually an existing lake, though some schemes dispense with the lower reservoir and pump their water from, and discharge their water to, a river. The upper reservoir might be an existing lake which may need enlarging or it might be entirely artificial. In the latter case a dam will have to be constructed high up on the mountain. It is an advantage if the lower reservoir is much larger than the upper one, so that the lower water level does not change too much. Large changes in level lead to an unsightly shoreline devoid of vegetation. The advantages of pumped storage are as follows: a very long lifetime, perhaps 70 years; a low running cost; a high efficiency, typically 75 per cent; and a short start-up time (less than 1 minute). The disadvantages are: a long lead time of 5 to 10 years; a high capital cost; and because the schemes are usually built far from load centres, high transmission costs. There may also be objections at the planning stage from environmental preservationists because, in the UK at least, the mountainous areas where such schemes could be built are also areas of outstanding natural beauty. However, several of the schemes already built have turned out to be tourist attractions in themselves.

There is a pumped storage plant at Ffestiniog in North Wales with an output of 360 MW and a storage capacity which is equivalent to full power for $3\frac{1}{2}$ hours. Full power can be brought up in under 1 minute. The head of water is 300 m and the same electrical machines serve as alternators and motors. This scheme was completed in 1963 and was the first pumped storage scheme in the UK.

A much larger pumped storage plant at Dinorwig near Llanberis in North Wales has a rated output of 1.7 GW. The storage capacity is equivalent to rated output for $4\frac{1}{2}$ hours.

The first pumped storage scheme in Scotland, which was built on Ben

Cruachan, has an output power of 400 MW. The scheme took advantage of a large existing lake, Loch Awe, as its lower reservoir. The upper reservoir, 350 m above, was formed by building a dam across the mouth of a natural corrie high up on the slopes of Ben Cruachan. The storage capacity is equivalent to full power for 21 hours. The reason for the large storage capacity is that the original idea was to work the plant on a weekly cycle, generating each day, pumping back rather less each night and then making up the difference by pumping for longer on Sunday. However, the plant is now used in a different way as explained below. The power station is deep inside the mountain, being at its lowest point 36 m below the level of Loch Awe. It houses four reversible Francis pump/turbines which are coupled to motor/alternators; that is to say, the same machine acts as either a pump or a turbine, and this is coupled to an electrical machine which can be used as a synchronous motor or an alternator. Loch Awe is so large, about 40 km^2, that pumping and generating have little effect on its water level. Even pumping for 21 hours would only lower Loch Awe by 20 cm.

Another pumped storage scheme in Scotland, at Foyers, uses the even larger Loch Ness as the lower reservoir. It has an output of 300MW, storage equivalent to full power for nearly 19 hours and a 175 m head of water.

There are three essentially different ways in which a pumped storage scheme can be used as part of the overall running of a power system. The first and original way was to work on a daily cycle, generating for several hours during times of peak demand and pumping during the early hours of the morning for long enough to refill the upper reservoir. Used in this way the load–duration curve is smoothed out as shown in Figure 7.6, with a consequent saving in running costs. The second way of operating is to switch from pumping to generating and from generating to pumping many times each day to balance the rest of the system against small changes in the demand. Cruachan is used in this way.

The third way of operating a pumped storage plant is to use it as a standby. The Dinorwig scheme was planned with this in mind and is designed so that it can be held available for immediate reserve generation for most of the day. In this mode the turbines are kept running at zero output by passing just enough water to overcome the losses and maintain synchronism with the grid. In this way the output can be raised to 1.32 GW in 10 seconds which will cover the sudden loss of two 660 MW steam turbine generators. Alternatively, the turbines can be kept running in compressed air, to hold back the water, using power from the grid at a lower operating cost. In this mode, the output can be raised to 1.32 GW in about 14 seconds.

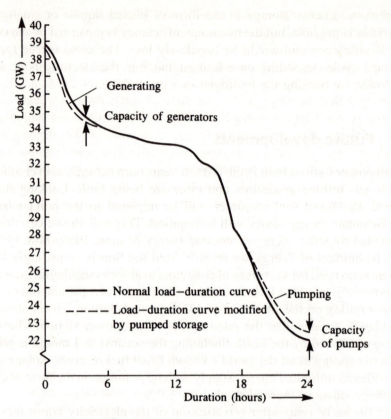

Figure 7.6 *Effect of pumped storage on a load–duration curve*

Ultimately the amount of pumped hydro storage is limited by the number of suitable sites, so other forms of energy storage are being investigated. Chemical batteries have already been mentioned. What the power system engineer needs is a new type of secondary cell which is cheap and long lasting. Electrical energy could be stored in capacitors. Storage densities of 3.6 MJ per litre are possible with thin glass-film dielectrics so a cube of 10 m sides would store 3.6 TJ, but storage life would be short owing to leakage and the energy would be very difficult to control. Energy can be stored in a rotating flywheel. A wheel of 4 m diameter, weighing, say, 100 tonnes, at 3500 rev/min would store about 36 GJ for short periods at a capital cost comparable with pumped storage. The efficiency would depend on frictional and windage losses and the length of time for which the energy was stored. Magnetic fields can be used to store energy by maintaining large currents in superconducting loops, but this system is likely to be very

expensive. Energy storage in the form of heated liquids or small solid particles is possible, but the maximum efficiency is governed by the steam cycle efficiency and would be hopelessly low. The same is true for any storage cycle depending on a heat engine, e.g. the electrolysis of water followed by burning the hydrogen in a gas turbine.

7.5 Future developments

Most power stations built in the last few years burn natural gas in combined-cycle gas turbine generators and more are being built. Looking further ahead, the fossil fuel resources will be depleted to the point where a replacement energy source will be required. This will almost certainly be provided by solar energy or nuclear energy or more likely both.

The amount of energy we receive from the Sun is surprisingly large. There is no need for vast areas of collectors to absorb meaningful quantities of power. Even in the UK, on a sunny day in summer, the amount of solar power falling on the site of a typical, large coal-burning power station at midday is greater than the rated output of the power station. The solar energy received by the Earth (including the oceans) in 1 month is greater than the energy in all the world's known fossil fuel reserves. Solar energy provides an unlimited energy supply with no pollution in the form of either waste products or thermal discharges.

In the early years after privatization of the electricity companies, the British Government imposed on consumers a fossil fuel levy of about 10 per cent of the cost of electricity. This levy was partly to fund the cost of decommissioning old nuclear power stations and partly to subsidize the development of renewable energy sources. Most work on using renewable energy, which in practice means solar energy, either direct or indirect, has been in four different technologies. The most extensive work has been on the use of wind-powered generators which were described in section 7.1.6. In contrast, a power station using direct sunlight has been built in the Mojave Desert, California, USA. It uses computer-controlled parabolic mirrors to focus sunlight onto tubes containing synthetic oil. The oil is heated to nearly 400 °C and then passed through heat exchangers to boil water and thus produce superheated steam which is used to drive steam turbine generators. The plant has a nominal capacity of 160 MW. Ocean waves contain energy which has been harnessed in a small way by experimental generators. The problems with wave energy are that very large structures are required to harness worthwhile quantities of energy, and they have to be strong enough to withstand the very worst weather; also the locations where wave energy is greatest are likely to be far from the grid. In many ways the most attractive way of using solar energy to generate electricity is provided by solar cells. These are used to power satellites in space and remote telephone exchanges. Solar cells have no moving parts and can be sited anywhere that receives sunlight, though the more sunny the location, the more electrical energy they will produce. Unfortunately they are prohibitively expensive for general use.

At present nuclear energy provides about 20 per cent of the electricity needs of England and Wales, but a much greater proportion in some countries.

Most nuclear power station reactors consume uranium-235 only. However, there is a way in which the latent energy in uranium-238 can be released. The process is indirect and involves a reactor known as a breeder reactor. Before explaining how this works, let us consider the ordinary natural uranium reactor in some detail.

When a uranium-235 atom undergoes induced fission the process can be described by the equation

$$^{235}_{92}U + n \rightarrow ^{236}_{92}U \rightarrow X + Y + (2 \text{ or } 3)n + \text{energy}$$

where n is a neutron and X and Y are fission products, typically elements with mass numbers near 95 and 140. The neutrons released have a high energy, they are 'fast' neutrons and must be slowed down to thermal energy before they can induce fission in another uranium-235 nucleus. The slowing down is done in the moderator which is placed between the fuel rods. In a natural uranium reactor the moderator is either graphite or heavy water (deuterium oxide). These materials slow down the neutrons without absorbing them significantly. Of the two or three neutrons released, one is needed to sustain the chain reaction while the remainder are absorbed elsewhere in the reactor. Some are absorbed in the fuel element canning material, some in the fission product nuclei, some in the control rods, some in the biological shield and some in the uranium-238. It is these last that are of interest. The process can be described by

$$^{238}_{92}U + n \rightarrow ^{239}_{92}U \xrightarrow{\beta} ^{239}_{93}Np \xrightarrow{\beta} ^{239}_{94}Pu$$

This means that a uranium-238 atom absorbs a neutron and becomes another isotope of uranium, uranium-239. This is radioactive and undergoes β (electron) emission with a 23.5 minute half-life, changing into neptunium-239. This in turn is also radioactive and decays by β

emission, with a half-life of 2.35 days, into plutonium-239. Plutonium-239 is a nuclear fuel with properties similar to uranium-235 in that it will undergo fission in a nuclear reactor to produce energy. Thus some uranium-238 is converted into fuel in a natural uranium reactor.

The processes occurring in the breeder reactor are similar to those described above, except that the geometry is rearranged to enhance the production of plutonium. The nuclear fuel in the core of a breeder reactor is plutonium-239 and the core is very small compared with a gas-cooled thermal reactor. There is no moderator, and fission is induced by fast neutrons. Hence these reactors are sometimes called fast reactors. The core is surrounded by a blanket of uranium-238 to absorb as many neutrons as possible and thus produce or 'breed' plutonium. With careful design it is possible to breed more plutonium in the blanket than is consumed in the core. Because of the small core size, the power density is very high. This makes the design of the cooling system very critical. There was a breeder reactor at Dounreay on the north coast of Scotland. This was cooled by liquid sodium, and although primarily experimental, it did deliver some power to the grid.

Breeder reactors can, at least in principle, utilize most of the available uranium and they can do even more. They can be designed to turn thorium, which does not undergo fission, into uranium-233 which is another fissile isotope of uranium. The process can be described by

$$ ^{232}_{90}\text{Th} + \text{n} \rightarrow\, ^{233}_{90}\text{Th} \xrightarrow{\beta} \,^{233}_{91}\text{Pa} \xrightarrow{\beta} \,^{233}_{92}\text{U}$$

The thorium atom after absorbing a neutron becomes the radioactive isotope thorium-233. This undergoes β emission with a half-life of 22 minutes, changing into protactinium-233. This in turn decays by β emission, with a half-life of 27 days, becoming uranium-233.

There are a number of major objections to the nuclear alternative. The two most serious of these are the risk of an accident to a reactor, releasing highly radioactive material into the atmosphere, and the problem of disposing of the radioactive fission products, some of which will remain significantly active for thousands of years. It is possible that future research will discover a way of de-activating the fission products, but there are no indications at present that a breakthrough is near.

A much more attractive nuclear energy approach to power generation lies in the use of nuclear fusion. This is the process which releases the enormous energy of the hydrogen bomb. Much research is being directed to the harnessing of fusion energy. One most promising reaction involves the fusion of two deuterium (heavy hydrogen) nuclei to form either helium or the even heavier isotope of hydrogen, tritium. The reactions are

$$_1^2H + _1^2H \rightarrow _1^3H + p + 0.64 \, pJ \text{ of energy}$$

or

$$_1^2H + _1^2H \rightarrow _2^3He + n + 0.51 \, pJ \text{ of energy}$$

where p is a proton and n is a neutron.

There are three very important advantages which these fusion processes have over the fission of uranium or plutonium, as a source of energy for electricity generation. First, there is an almost inexhaustible supply of deuterium in the sea. Second, there is no large accumulation of radioactive material in the power station, which might be released in the event of an accident. Third, the only radioactive by-product is tritium which could easily be dealt with and might perhaps be used in the even more energetic reaction

$$_1^2H + _1^3H \rightarrow _2^4He + n + 2.8 \, pJ \text{ of energy}$$

It may well be that one or more of these three reactions will supply much of our energy requirements in the 21st century. It is also possible that the major energy source will be the Sun and that micro-organisms or plants, specially developed by the techniques of genetic engineering, will convert sunlight, water and atmospheric carbon dioxide into alcohol or methane, or perhaps hydrogen.

7.6 Summary

Almost all the world's electrical energy is generated from fossil fuel, nuclear energy and solar energy. The important fossil fuels are coal (including lignite and peat), oil and natural gas. The main source of nuclear energy is uranium-235. The important applications of solar energy, in both cases indirect, are hydro-electric and wind power. The different energy sources have different costs and require different technologies to turn them into electricity. The costs can be divided into the running cost, mainly the cost of the fuel itself, and the capital cost of building a power plant. Often low or zero fuel costs, e.g. nuclear and hydro-electric power, are associated with high capital costs.

Electricity cannot be stored in worthwhile quantities, so a power system has to be managed so that the amount of electricity generated always closely matches the demand. The demand varies with the time of day, the seasons and the consumption patterns of consumers. The base load, which must be met for 24 hours each day, is supplied by the stations which have the lowest running costs. In contrast, the peak load can be met by stations which have a high fuel cost, provided they have a low capital cost or low capital value. Speed of response to changes in demand is also important for generators supplying the peak load.

Energy can be stored at times of low demand, by using hydro-electric pumped storage plants. These plants can then generate at times of peak demand in the same way as a normal hydro-electric station.

When reserves of fossil fuels, particularly natural gas and oil, are depleted, the future is likely to see increased use of solar energy and nuclear energy.

Further reading

More advanced books on power systems:

Glover, D. G. and Sarma, M. (1987) *Power System Analysis and Design*, PWS, Boston.

Grainger, J. J. and Stevenson, W. D. (1994) *Power System Analysis*, McGraw-Hill, New York.

Nasar, S. (1990) *Schaum's Outline of Electric Power Systems*, McGraw-Hill, New York.

Stevenson, W. D. (1982) *Elements of Power System Analysis*, McGraw-Hill, New York.

Weedy, B. M. (1987) *Electric Power Systems*, Wiley, Chichester.

Books on specific aspects of power systems:

Berrie, T. W. (1983) *Power System Economics*, Peter Peregrinus for IEE, Stevenage.

Burgoyne, D. J. (1985) *The Dinorwig Power Station*, Mechanical Engineering Publications, Bury St Edmonds.

Chard, F. de la C. (1976) *Electricity Supply: Transmission and Distribution*, Longman, Harlow.

Culp, A. W. (1991) *Principles of Energy Conversion*, McGraw-Hill, New York.

Flurscheim, C. H. (ed.) (1982) *Power Circuit Breaker Theory and Design*, Peter Peregrinus for IEE, Stevenage.

Freeman, P. J. (1974) *Electric Power Transmission and Distribution*, Harrap, London.

Freris, L. L. (1990) *Wind Energy Conversion Systems*, Prentice Hall, Hemel Hempstead.

The Institution of Civil Engineers (1987) *Tidal Power*, Proceedings of the Symposium held in London, 30–31 October 1986, Thomas Telford, London.

Jefferies, D. (1992) *Experience with Restructuring and Privatising Electricity in England and Wales*, GIGRE Keynote Address, National Grid Company, Coventry.

Kessler, G. (1983) *Nuclear Fission Reactors*, Springer-Verlag, Vienna.

Lewis, E. E. (1977) *Nuclear Power Reactor Safety*, Wiley, New York.

Osborne, J. (ed.) (1994) *International Water Power and Dam Construction Handbook 1995*, Reed, London.

Taylor, E. O. and Boal, G. A. (eds) (1969) *Power System Economics*, Arnold, London.

Index